William Ford Robinson Stanley

A descriptive treatise on mathematical drawing instruments

William Ford Robinson Stanley

A descriptive treatise on mathematical drawing instruments

ISBN/EAN: 9783337276225

Printed in Europe, USA, Canada, Australia, Japan

Cover: Foto ©berggeist007 / pixelio.de

More available books at **www.hansebooks.com**

A DESCRIPTIVE TREATISE

ON

MATHEMATICAL

RAWING INSTRUMENTS,

THEIR

CONSTRUCTION, USES, QUALITIES, SELECTION,
ISERVATION, AND SUGGESTIONS FOR IMPROVEMENTS.

WITH

Hints upon Drawing and Colouring.

BY

WILLIAM FORD STANLEY, M.R.I.,

*athematical Instrument Maker to H.M. Government, Science and Art Department, Council
of India, Admiralty, Tithe Commission Office, Royal School of Naval Architecture,
Royal Military Academy, Royal Geological Society, etc., etc.*

"C'était la main de l'homme qui était la seule machine de l'esprit."
A. DE LAMARTINE.

FIFTH EDITION.

PUBLISHED BY
. & F. N. SPON, 46, CHARING CROSS.
NEW YORK: 446, BROOME STREET;
AND
HE AUTHOR, AT 5, GREAT TURNSTILE HOLBORN, LONDON, W.C
1878.
Price Five Shillings.

BUTLER & TANNER,
THE SELWOOD PRINTING WORKS,
FROME, AND LONDON.

PREFACE,
TO THE FIRST EDITION.

OUR scientific literature has become so diffuse and universal, that we expect to find an outline of all the little mysteries of any particular art, somewhere in print. If such is not to be found, the restless inquiry that demand proclaims, will generally tempt some individual to become the teacher, however far he may fall short of perfect mastery.

Certainly, the author of the following pages, who could only spare desultory hours from active business, would not have attempted to write a description of the qualities and uses of Mathematical Drawing Instruments,—being conscious that his powers were greater with the lathe and file than in the "ways of gentle rhetoric,"—had he not felt that there was really a want of such a work, much of his time being constantly required to describe by letter instruments which, from their extensive use by some of the profession, ought to be known of at least by all : this particularly applies to such instruments as the Eidograph, Centrolinead, Computing Scale, and some others of like utility. Further, it appears so much more English to purchase complete information on any subject *if you can,* than to be compelled to ask *small* particulars of any person in detail.

It happened that a treatise upon Mathematical Instruments really must be *written,* to be produced at all; scissors could do no more. The ignorance of the mere compiler, in this line, had become so striking as to be only ridiculous. Here is an instance. A very silly triangular compass, which consisted of three jointed arms, moveable upon a horizontal centre, was illustrated in a work upon mathematical instruments published over a hundred years since. It is not very certain whether the instrument was ever made; but the next writer who wrote upon the subject extracted the description, also the engraving, except that his engraver, accidentally no doubt, made the joint very small. The next writer who recommended the instrument left out the joint altogether, whereby it ceased to be a triangular compass, except in the faint *historical* similarity to the original. Subsequent writers, finding the stereotype to hand, unfortunately followed the *last* description, and also the *last* engraving, very much to the perplexity of the more philosophical reader.

It is not intended that the above should infer that we have no original works upon mathematical instruments. We have several—but they are of the far past.

It will be attempted in the following pages to review the merits of a few of the possibly useful instruments to be found in them.

Really the best work we have is the "Geometrical and Graphical Essays" by George Adams, published in 1791. This was a rather complete work in its day. It embraced some description of all the instruments then in use. It was practical too,—written by a workman and a shopkeeper in constant intercourse with the user. There is one deficiency which the writer appears to have felt—that of his not being a draughtsman. He made, however, an excellent apology by offering very copious opinions of professional men.

Of more modern works, the only one of this class with any claim to originality is a "Treatise on the Principal Mathematical Instruments," by F. W. Sims, 1844. This is practical one way; the writer is a professional draughtsman. The work, however, is limited to instruments for the use of the land surveyor; and of these there is the omission of some of the most important.

After consideration of all that has been done, the writer of these pages determined to place before his readers only his own opinions and experiences of drawing instruments. The plan is egotistical, but it offers the reader all the writer *really knows* on the subject, with perhaps some of his fancies and ideas, described of

PREFACE TO THE FIFTH EDITION.

SOME few matters are added in this edition, and some further remarks on principles of construction and professional values and uses of certain instruments; also, some further hints on drawing and colouring. The conditions under which the first edition was produced, after twelve years, no longer remain. This Treatise, independent of its steady moderate sale, has also met with such an amount of literary patronage, that its contents may now be found, in various stages of dilution, widely diffused in works that touch upon its subject. The old stereotype of imaginary triangular compasses described in the first Preface, has been put by. Instruments, and ideas of them, introduced for the first time in these pages, have been very generally accepted by the profession, and the patterns and principles imitated by the trade. Therefore the first Preface could now very well be omitted, but it is left intact, as being quite true at the time it was written; and it may be now used as a mark of our progress in small things. This change of circumstances calls forth these few supplementary remarks.

GREAT TURNSTILE,
  Aug., 1878.

CONTENTS.

SECTION I.

Relates to Drawing Instruments used to produce Lines and Geometrical Figures, some of which are also used to set off Spaces.

CHAPTER VIII.

CHAPTER IX.

CHAPTER X.

CHAPTER XI.

CHAPTER XII.

CHAPTER XIII.

CHAPTER XIV.

CHAPTER XV.

CHAPTER XVI.

CHAPTER XVII.

SECTION II.

Relates to Drawing Instruments used as Guiding Edges, Instruments for Measuring, and Drawing Materials.

MATHEMATICAL
DRAWING INSTRUMENTS.

SECTION I.

*Relates to Drawing Instruments used to produce Lines
and Geometrical Figures, some of which are also used
to set off Spaces.*

CHAPTER I.

INTRODUCTION—ARRANGEMENT OF INSTRUMENTS—DEFINITIONS—QUALITIES AND FINISH.

THIS chapter is devoted to desultory matters relating
to drawing instruments generally, and is intended
to introduce and unite the subject consistently. The
plan of the work to be followed in all future chapters will be to separate each subject, by placing all
the relative instruments, or those intended to produce
like results or like forms, separately in consecutive
chapters.

As a case of drawing instruments generally embraces
our first ideas of mathematical instruments, and at the
present time belongs almost as much to our school requirements as the slate and pencil, it may be well, by
way of introduction, to give a slight technical description of these cases as they are arranged for professional
purposes, particularly as the instruments contained in a
case form a collection of those most useful and import-

B

ant, and therefore should not be lost sight of. The simplest cases contain the most necessary instruments, and the more expensive and complete contain what may be termed the draughtsman's luxuries.

If we presume the reader to be acquainted with the names of the ordinary drawing instruments, the cases of instruments, as they are technically named, are the *Half-set Case*, which contains a pair of compasses with movable leg, ink point, pencil point and lengthening bar, and drawing pen; the *Set Case*, which contains the same as the half-set case, and in addition, a pair of dividers, ink bow, and pencil bow; the *Full-set Case*, which contains, in addition to the instruments in the set case, a set of spring bows, a pricker, and one extra drawing pen: this last case, containing all the instruments constantly required, is sufficiently complete for ordinary professional purposes. Cases with a greater number of instruments, containing proportional compasses, road pen, wheel pen, tracer, beam compasses, etc., are termed *Long-set Cases*, which term is indefinite as to the quantity of instruments. The above cases also generally contain three rules of very little use—a protractor, sector, and parallel.

The above is the general arrangement of instruments in cases, which is however sometimes varied; as, for instance, tubular compasses may be put in the place of the half-set, which will answer in practice for the same purposes; or other changes may be made, to the taste of the draughtsman.

Persons with limited means will find it better to procure good instruments separately of any respectable maker, as they may be able to afford them, than to purchase a complete set of *inferior* instruments in a case. With an idea of economy, some will purchase second-hand instruments, which generally leads to disappointment, from the fact that inferior instruments are manufactured upon a large scale, purposely to be sold as second-hand to purchasers, principally from the country, who are frequently both unacquainted with the workmanship of the instruments and of the system practised.

Inferior instruments will never wear satisfactorily, whereas those well made improve by use, and attain a peculiar working smoothness. The extra cost of purchasing the case and the nearly useless rules would, in many instances, be equal to the difference between a good and an inferior set of instruments without the case. Further, if the case be dispensed with for economy the instruments may be carefully preserved by merely rolling them up in a piece of wash leather, leaving space between them that they may not rub each other ; or, what is better, by having some loops sewn on the leather to slip each instrument separately under.

Before leaving the subject of cases of instruments, a few words may be said upon the various kinds of cases made to contain drawing instruments, as it would be difficult to return to the subject in an advanced part of the work.

The cases in most common use are made seven inches long, by from four to six inches wide, and about one inch and a half deep ; they are generally made of mahogany, frequently veneered with rosewood or walnut. They are much better if made of solid wood, with dovetailed corners, as the veneers, although very pretty at first, through the necessity of laying them upon the plain wood, by soaking them with glue, slowly contract and very generally draw the tops hollow and the sides out of square. Oak is a very suitable material, looks nice, and stands well.

The cases that are found practically most convenient are made thirteen inches long; these are termed magazine cases, and will contain 12-inch scales, angles, curves, and other useful instruments. They are generally made in an elegant and costly manner, veneered, with metal bands and capped corners; they may, however, be made plainly in the solid, at a very moderate cost. They possess one advantage over the 7-inch cases,— that is, they will contain nearly all the necessary drawing materials, and will be found, upon the whole, cheaper to the professional draughtsman than the cost of several separate boxes.

Many draughtsmen, for the convenience of having their instruments at hand when required, prefer a pocket-case. This is a thin wood case, covered with Russia or Morocco leather, and is generally made to contain the set or full set of instruments. Pocket-cases should be made with the corners properly rounded; the fastening should be a spring clip or a bolt, as adopted by the French: we often see them fastened by hooks, which catch in everything.

The French make pocket-cases very tastily; they wear much better and are thinner than the old style of English ones, the sides and corners being entirely rounded. Lately the English case-makers have imitated the French, nearly equalling them in appearance, and perhaps surpassing them in the solidity of the leather work and hinging.

Such instruments as are generally included in the ordinary cases are made of *two* sizes only, which are called 6-*inch sets* and 4½-*inch sets;* one of these terms is applied to the whole set, whichever it may be, although only the compasses and drawing pen are of the length from which the set is named. The ink and pencil bows of a 6-inch set are only three inches long, but are called 6-inch bows. The same rule is, for convenience, applied to all the instruments, the term indicating *in proportion with* 6-inch or 4½-inch compasses. Six-inch sets of instruments are best suited and most used by mechanical engineers. Four-and-a-half-inch sets are placed in pocket-cases exclusively, and are used almost entirely by architects, and civil engineers.

The METALS generally used in making drawing instruments are brass, electrum, or silver, the points and joints being made of steel. Silver is little used of late years, being costly and possessing little merit over electrum, which is at present the most popular. Electrum, *as it is called,* should be an alloy composed of pure nickel and copper; in colour it should nearly equal the whiteness of standard silver, with the advantage of being stiffer and of less specific gravity. Its merit over brass is that it will not soil the fingers by forming verdigris from the action of the perspiration of the hand; neither does it emit the odour peculiar to brass, which is very disagreeable to some few sensitive persons. Electrum of inferior quality approaches the colour of pale brass.

Attempts have been made to introduce aluminium and its alloy, aluminium-bronze, into the manufacture of drawing-instruments—it must be admitted, with little success.   Aluminium would undoubtedly in some respects be very excellent, especially for its non-corrosive quality and extreme lightness; but it requires some genius to discover the method of soldering it to steel, to render it at all adapted to drawing instruments.   With regard to aluminium-bronze, after many experiments, the writer has been unable to discover any peculiar merit that it possesses over brass.   As the steel of instruments is very liable to rust, no greater part of the instrument should be made of that metal than is necessary for strength or fineness of edge.

In concluding these remarks, which apply to drawing instruments generally, a few observations may be made on what is technically termed *finish*, as it is this which marks the taste of the workman, and is considered a test of good work; so that many professional gentlemen select their instruments by observation of the finish only, relying upon the work being well done if pains have been taken in this particular.

By finish is understood the grace and correspondence in form of each side and opposite part of the instrument, and of the quality of the surfaces, which should be perfectly flat, or straight, in one direction, and show equally sharp angles.   The grain left from finishing the surface should be at right angles with the length of the instrument; it should in all parts show an equal and very fine-grained surface, but not a burnished gloss.

The French and Swiss instruments were very popular some few years since, but seem to have gone out of favour; they appear to have owed their popularity to their glossy, burnished surfaces, easily produced by the buff wheel.   It is creditable to the English workman that he never adopted this fashion, although it was making inroads upon his trade; had he done so, it would have degraded his work of "*hand and eye*" to the rank of ironmongery. It must, however, in justice be remarked that although the French and Swiss ordinary cheap work, such as is commonly sent to this country, is highly polished, their better class of work is hand-finished, as our own; and in such instruments as the theodolite and microscope they leave the *grain* more distinct than it is left by our English workmen.

CHAPTER II.

INSTRUMENTS FOR PRODUCING FINE LINES—DRAWING PENS.

THE DRAWING PEN is perhaps the most important instrument to the draughtsman, being used to render nearly all the lines of a drawing permanent with Indian ink after the first outline has been produced with the blacklead pencil. The general construction of the drawing pen is perhaps too well known to need particular description. It consists of two pointed blades of metal, which are fixed or jointed over each other in such a manner as to leave a space sufficient to support the ink by capillary attraction. The distance of the points is adjusted by a milled-head screw ; the line to be produced by the pen corresponding in thickness with the adjustment. Drawing pens are differently constructed, both for economy and convenience ; each of the various kinds being frequently adapted to a special purpose.

Fine Drawing Pen.

The FINE DRAWING PEN is cut out of a piece of steel wire, the whole working part of the pen being in one piece. It is generally preferred for fine-line drawing, from the fact that the nibs are each equally firm when in contact with the drawing-paper; this is an important consideration, difficult to attain with any form of jointed pen. The fine pen is much used for plotting surveys; it is lighter than the jointed pen and less angular, thereby turning more readily in the fingers to follow irregular lines. The fault of the pen is the difficulty of cleaning between the nibs, and of setting it. For these reasons one handle is frequently adapted to six pens, for the convenience of draughtsmen having them all set at one time by the maker. The setting is undoubtedly a difficulty, but pens will retain their working condition for a considerable time if they be kept perfectly clean. The light coating of rust which occasionally accumulates between the nibs may be easily removed by folding a narrow piece of No. 0 glass paper, and drawing it between the nibs until they appear bright; this should be done without touching the extreme points. It will also be found to keep the pen in good working order.

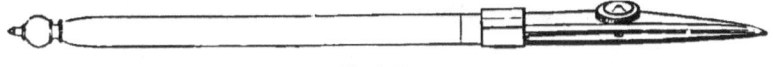

Block Pen.

The BLOCK PEN is similar to the fine pen, but the blades which form the nibs are soldered into a block of metal, instead of being sawn out of a piece of solid steel; the nibs are also much wider than those of the fine pen, and consequently support a larger supply of ink. This, which is the least expensive large pen, is useful for making working or detail drawings.

The LITHOGRAPHIC PEN, used for drawing on stone, is similar to the fine drawing pen, only that it is much larger and longer in the nibs. The steel of this pen should be hardened to brittleness.

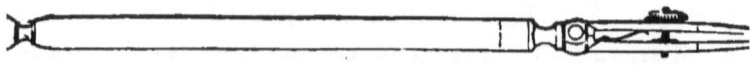

Lifting-nib Pen.

The LIFTING-NIB or JOINTED PEN, which is the one mostly used, is made only partly of steel, the body of the pen being generally of electrum, which keeps cleaner in use than steel does. The upper nib is hinged to lift entirely, so that the inside may be easily cleaned or kept clean, an important advantage, as no pen will draw a neat, continuous line, however well set, if the inside is either corroded or clogged with ink or rust. The upper nib in the lifting-nib pen is forced open by a small spring, which causes it to follow the adjustment of the screw to produce the required thickness of line. The defect of this principle is that it is scarcely possible to make the joint sufficiently perfect for the upper nib to be equally firm with the back nib, if the joint is loose enough for the small spring to lift it; if it is not so made, the back nib takes the greater pressure, and consequently scratches,—a fault common to this construction.

Improved Drawing Pen.

The writer has improved the construction of the lifting-nib drawing pen, obviating the fault of the loose joint by making the lifting-nib entirely of steel, and so constructed that it forms a lifting-spring in itself. By this means the small spring between the nibs is dispensed with, and the joint may be made perfectly tight; therefore the pen cannot scratch if properly set. There is another equally common fault in drawing pens, that of the nibs being so fine and weak that they partially close, and produce unequal lines by the pressure necessary to be used against the guiding edge. In this improved pen the defect is remedied by making the back nib strong enough to resist the pressure, at the same time leaving the upper nib sufficiently thin to adjust easily with the screw. A similar construction of pen may be made without the lifting-nib.

There is a kind of drawing pen, very little used at the present time, which is made entirely of steel, the handle being produced by a continuation of the nibs. This pen opens by a joint in the centre *sideways,* in the manner of a pair of scissors, after removing the adjusting screw and a small cap on the top. It would be a good pen were it not for its weight, and the liability of the handle to rust.

Curve Pen.

The writer, some time ago, invented a peculiar pen, which much pleased some few of the profession, who have since used no other. This has a crank, or arm, above the pen, as shown in the engraving. The crank is intended to prevent the interference of the hand with the observation of the nibs, thus enabling the pen to

be held quite upright; it also possesses the advantage
that it will by a slight twist of the fingers follow any
straight or curved edge with steady motion.   Its faults
are, that if the pen be made sufficiently large, it will
take the hand too far off the drawing to be convenient; if it be made small, it will allow of but a
scanty supply of ink.

In using a drawing pen of any kind it should be
held very nearly upright, between the thumb and first
and second fingers, the knuckles being bent, so that it
may be held at right angles with the length of the
hand.   The handle should incline only a very little, say
ten degrees.   No ink should be used except Indian ink,
which should be rubbed up fresh every day upon a
clean palette, of one of the kinds described farther on.
Liquid ink and other similar preparations are generally
failures.   The ink should be moderately thick, so that
the pen when slightly shaken will retain it a fifth of
an inch up the nibs.   The pen is supplied by breathing
between the nibs before immersion in the ink, or by
means of a small camel-hair brush; the nibs will afterwards require to be wiped, to prevent the ink going
upon the edge of the instrument to be drawn against.
The edge used to direct the pen should in no instance
be of less than a sixteenth of an inch in thickness; a
fourteenth of an inch is perhaps the best.   If the edge
be very thin, it is almost impossible to prevent the ink
escaping upon it, with the great risk of its getting
on to the drawing.   Before putting the pen away, it
should be carefully wiped between the nibs by drawing a piece of folded paper through them until they
are dry and clean.

After considerable wear, the drawing-pen will require

setting. This is done by sharpening the nibs on an oil-stone; the kind of stone known as *Arkansas* answers the purpose best. The operation of setting requires considerable judgment and practice, and is one of those mechanical niceties which it is difficult to describe. It will generally be found better to have the pen set by a respectable instrument-maker, where one can be found within a convenient distance, than for an inexperienced person to attempt it. The best information that can be offered upon the subject is to describe how the pen should appear when it is properly set, which may be some guide to the operation, and may be given in a few words. The ends of the nib should be alike and equally round, in form of the top half of the letter o in this type, and the edges of the nibs should appear equally thin when held to the light, but not so sharp as to scratch the thumb-nail when they are drawn across it.

Lithographic Crow-quill.

Besides the mathematical drawing pens described in this chapter, in modern practice, for plotting irregular outlines on plans, or drawing-in small ornaments in architectural elevation, a very fine kind of writing-pen, termed a *mapping pen*, or another kind termed a *lithographic crow-quill*, is commonly employed; either of which is rather more convenient to use for these purposes than any description of mathematical pen.

CHAPTER III.

INSTRUMENTS FOR PRODUCING BROAD LINES, DOUBLE LINES, DOTTED LINES, TRANSFER, ETC.—BORDERING PEN—ROAD PEN AND PENCIL—WHEEL PEN—TRACER—PRICKER.

THE BORDERING PEN is used for producing very broad lines, applied principally to border drawings. The ordinary construction of this pen is similar to the block pen already described, except that the nibs and points are considerably wider, and very slightly bowed; the nibs are also closer, the distance at the adjusting screw not being much over the sixteenth of an inch; by this construction the pen supports a large amount of ink by capillary attraction. The set of the pen should not be nearly so sharp as that of the ordinary drawing pen.

The author has found a great improvement may be made in the bordering pen by making an extra inner nib, as shown in the illustration below. A pen so

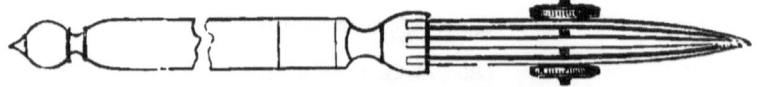

Improved Bordering and Colouring Pen.

constructed will contain double the quantity of ink to that of the ordinary form, and will if necessary, produce a line of double the thickness. The adjustment to width of line is obtained by two screws, which adjust from each side of the pen. The inner nib is made a little shorter than the outer ones, as it is not necessary that it should touch the drawing. Besides the ordinary purpose of the bordering pen, it may be used for colouring the narrow lines frequently required upon drawings, as for roof timbers, partition walls, iron rods, shadow lines, etc. For these purposes it is better if made entirely of electrum, which is less corrosive than steel, and in other respects answers equally well.

Road Pen.

The ROAD PEN is used for producing continuous parallel lines, particularly where many are required at equal distances, as roads, and hedge lines upon surveys, railway lines, joists, etc. In construction it consists of two drawing pens united to the handle by flat springs, which are connected together by an adjusting screw to enable the pens to be set at any required distance apart. The road pen is generally made entirely out of one piece of steel, which should be of sufficient thickness to make the nibs wide enough to hold a fair supply of ink. The spring sides should be made moderately long, that the nibs of the pen, when nearly closed, should incline as little as possible. When this pen is properly set, the four nibs, if screwed together, will appear as one line.

Section Pen.

There is a kind of pen which is not much known, but which may be conveniently used for drawing sectional lines upon mechanical drawings, or other parallels at equal distances. It resembles the road pen just described in general appearance, except that in place of one of the pens there is a plain flat-pointed nib, which is made a little shorter than the drawing pen. If this nib be set to the required distance of the section lines of a drawing, it may be placed consecutively upon the last line drawn, and produce an equal sectional tint.

Road Pencil.

The Road Pencil is similar to the road pen, except that it has two spring sockets fitted to carry pencils in place of pens. It is used for the same purpose as the road pen, and is a very convenient instrument for architectural drawing, as it may be set to the thickness of wall, joist, or other constant quantities, and produce the pair of lines perfectly. It may be frequently used where the road pen could not, as neatness in the beginning and finishing of pencil lines is not so absolutely requisite as it is for the finished ink lines.

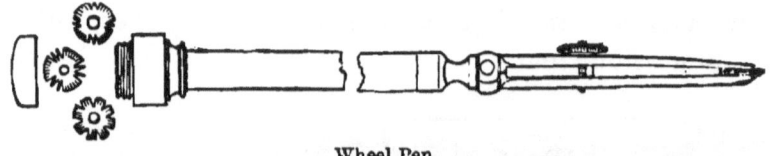

Wheel Pen.

The Dotting or Wheel Pen consists of two nibs pointed together in the manner of the drawing pen. At the point of one of the nibs is a fixed pin, upon which is placed one of the rowels or wheels represented

in the illustration; the upper or lifting nib is afterwards closed upon the rowel, which retains it in its place.

The dotting pen is used for producing dotted lines for boundaries, proposed works, hidden parts of architectural or mechanical drawings, etc. It requires very great care in using, and works generally better if it is not held quite straight with the line to be produced. The Indian ink should be mixed thicker than for the ordinary drawing pen. Before producing any line, it is necessary to run the pen over a piece of waste paper, and to observe whether it is properly charged with ink; otherwise it frequently happens that the ink will be entirely carried down by the rowel, and spoil the drawing. For this reason the dotting pen has been entirely abandoned by many draughtsmen, although a very useful instrument when made to act properly. The failing in the ordinary forms of dotting pens is the insufficient capillary attraction. There have been many schemes to remedy this defect, both in this country and abroad. The French method is to place a small brush above the rowel; this is not very effective; it will not supply properly, and is liable to splutter without it is used with great care. The Swiss place a large ivory wheel above the rowel which turns with it; this plan does not ensure regular dotting.

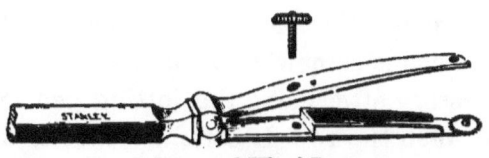

Improved Wheel Pen.

The best method is to make the ordinary dotting pen rather close in the nibs, and to enclose the open

space between the nibs by surrounding the under nib
with a thin web of metal, thus forming a narrow
chamber for the ink, which, when the upper nib is
down, will be closed on all sides nearly to the top of
the rowel.   The greater capillary attraction caused by
the four walls of this small chamber effectually with-
holds the ink from running down.   The ink is supplied
to the chamber with a camel-hair brush, the upper nib
requiring to be raised to open a small crevice, across
which the brush, charged with ink, may be drawn.
After this pen is charged, some pains are required in
commencing to use it; it is well always to try it on a
piece of waste paper, as before mentioned, particularly
to ascertain if the ink be of proper consistence.   After
getting it in working order, it will produce several
yards of dotted lines without defect.

The rowels are made of different patterns, the most
common being made to produce the following lines.

The first is used for hidden parts of mechanical or
architectural drawings, lines over which the chain
passed in surveying, or of imaginary lines of measurement.   The second is used for boundaries of a township.
The third for boundaries of a parish.   The fourth and
fifth for proposed railways, canals, roads, or similar
works.   Although these are very general applications,
their special use is somewhat arbitrary.

It is necessary after using the dotting pen to clean it
out, and to wash the rowels carefully, also to dry them

quickly with blotting-paper, as they are very liable to corrode with rust or ink.

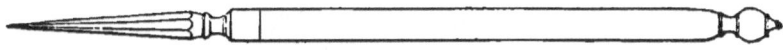

Tracer.

The TRACER, illustrated above, is a tapering point of hard steel, fixed into an ivory handle. The point is slightly rounded, so that it may be drawn across a sheet of paper with considerable pressure without scratching. It is used for copying drawings by pressing firmly with the point over the lines to be reproduced. The drawing is placed over the sheet of paper which is to receive the copy, with a sheet of black-lead or carbonic paper interposed. The tracer is mostly used to reproduce a drawing from a tracing, often to make a finished copy, or to avoid errors or small alterations in the first drawing. It is also used for etching on metal. Tracers are sometimes made of agate, which is harder, smoother, and altogether better for the purpose than steel.

Pricker.

The PRICKER or needle holder is used for marking off distances by scale, also for copying by pricking through all the angles or important parts of a drawing to one or more sheets of paper beneath, which are intended for copies, the lines being produced by connecting the points after the original is removed. It is one of the instruments generally supplied in a full-set case, and consists of a common needle held by some mechanical contrivance. The usual manner of holding the needle is that illustrated above: the needle is placed between two jaws which have a small v groove up the centre, inside of each; the jaws are slightly bowed, so that a slide ring passed over them may press them tightly together and secure the needle. Although this is the form commonly used, it is by no means effective; the needle being held only by the spring of the jaws, it frequently slips back, if it is not of exactly the length to be stopped by the bottom of the space between them. It also pulls out if required to prick through several thicknesses of paper. There are many expedients intended to remedy these faults; one is by making the point similar to the ordinary needle-holding compass-point. This plan is little better than the other, as the needle slips as readily, only that it is a little more easily replaced. Another method is by fixing the needle in a point similar to an ever-pointed pencil: this method is complicated and not effective.

Patent Needle Holder.

The author patented a manner of holding needles which he finds universally approved of. This is effected by inserting a bolt through the side of the instrument, the bolt having a hole in the head, through which the needle is passed; a milled nut on the bolt draws the needle tightly against the side of the instrument; the point of the needle is passed through a hole in the point of the instrument, which it exactly fits. This needle holder is very effective, and the needle is easily replaced if broken.

CHAPTER IV.

INSTRUMENTS FOR DIVIDING AND MARKING OFF DIS-
TANCES—DIVIDERS—DESCRIPTION OF POINTS, ETC.
—HAIR DIVIDERS—PORTABLE DIVIDERS, ETC.

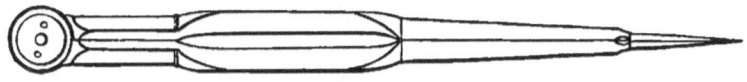

Plain Dividers.

THE PLAIN DIVIDERS are the most simple form of compasses; they are too well known to need further description than that conveyed by the illustration above.
They are employed to divide spaces into equal parts, to
set off or transfer distances, and to copy drawings. It
was formerly the practice to set off all distances from a
scale by a pair of dividers, the scales being divided up
the central portion only of a piece of boxwood or ivory.
The present practice is to set off all distances from the
edge of a scale with a pencil or pricker, except small
distances that require repeating on different parts of
the drawing. For this reason the dividers are less used
than formerly. Nevertheless they are a very important
instrument for the purposes to which they can be
advantageously applied.

In the selection of a pair of dividers there are many
qualities to be observed; the points should be sharp
and as fine as possible compatible with steadiness, but
not too light, or they will be springy and difficult to
divide with. The points when closed should fall exactly
over each other, forming as it were one point. But

above all, the dividers should possess a perfect joint. There are two common methods of making the joints, which it will be necessary to describe fully, as the same remarks will apply to every description of compasses.

The joints of compasses in all instances should be of two metals; the one of the material of which the compasses are made—generally brass or electrum, the other of steel. This is necessary, as two pieces of the same kind of metal, except hardened steel, when moved in close contact, abrade their own surfaces, or what is technically called *fret*. There should always be two steel plates to every joint; if one plate only is used, as in the Swiss and French instruments, the centre on which the joint works has to be screwed very tightly, to produce the necessary friction, and this causes rapid wear. With two plates it is not necessary to screw the centre sufficiently tight to cause appreciable wear, and any imperfection of workmanship in the fitting of one of the plates is partially counteracted in the other.

Long Joint.   Sector Joint

In what is technically called the *Long Joint*, which is the oldest form applied to compasses, and which is still used almost exclusively on the Continent, the joint extends some distance down the body, as shown in first illustration above; consequently, with the closing of the compasses, a larger amount of surface comes continually into action, producing much greater friction and stiffness when the instrument is nearly closed than when it is wide open. This is the fault of the joint; its only merit is that it requires little skill in making, as it admits of fitting up, if the work has been commenced improperly. It is universally used for common instruments, for which it answers very well if properly made.

The other form of joint in use, technically called the *Sector Joint,* is now made to all compasses with any pretence to good quality of workmanship. In this joint the working surfaces are of circular form, equally distributed around the centre; consequently the compasses move with equal pressure, whether nearly closed or wide open. It is not necessary to screw the centres of sector-jointed compasses tightly, as the surfaces are not required to come in *perfect* contact—for this reason, that after the sector joint is made, the workman lubricates between the joint with hot bees-wax, which aids in producing that peculiar deadness in movement so much esteemed in the sector joint. If the joint be made true, the wax will never leave it. The writer has parted sector joints which have had twenty years of constant wear: the wax appeared the same in quantity as when first introduced, although it was blackened, the amount of wear on the plates not having been sufficient to take out the fine file marks which were left in making the joint. This is a peculiarity of the sector joint which no other possesses.

The sector joint to attain perfect movement should be tightened and loosened as little as possible, thus allowing it to form its own surface. Care should be taken that no oil gets upon the joint, as it speedily dissolves the wax and spoils the joint; and young draughtsmen should not be trusted with a key to tighten or loosen the joint, as, if it is properly adjusted, it need not be disturbed for four or five years.

The method of trying if a joint is perfect is to open the compasses until they are in line, and then to close them again very slowly, noticing if equal pressure is required at all openings: this will test the evenness of the joint. Another important consideration is, that the centre should fit perfectly; to examine this it will be necessary to take the compasses about half open, and close and open them alternately and quickly for as small a distance as possible, as it were to feel the joint. In doing this, if the centre should not fit, a slight jerk will be felt immediately after commencing to open or close them. Improperly made sector joints are worse to work with than improperly made long joints.

The following hints for using dividers or compasses may be useful to the young draughtsman. It is considered best to place the forefinger upon the head, and to move the legs with the second finger and thumb. In dividing distances into equal parts, called *stepping*, it is best to hold the dividers as much as possible by the head joint, after they are set to the required dimensions, as by touching the legs they are liable to change if the joint moves softly, as it should. In dividing a line, it is better to move the dividers alternately above and below the line from each point of division, than to roll them over continually in one direction, as it saves the shifting of the fingers on the head of the dividers. In taking off distances with dividers, it is always better first to open them a little too wide, and afterwards close them to the point required, than to attempt to set them by opening.

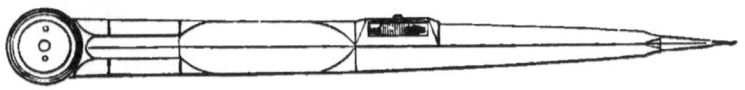

Hair Dividers, closed.

The HAIR DIVIDERS differ from the plain dividers in having one of the points moveable for a short distance by a screw adjustment, which enables the draughtsman to adjust the point with greater delicacy than could be attained by moving the head joint only. The general form of fine adjustment consists in the upper portion of one of the points being formed into a spring, which is sunk into a groove up the inside of one leg of the dividers. The inclination of the spring is to move inwards to close the points, but it is retained and adjusted as required, by a milled-head screw. Instead of a screw, a nut sunk into a mortise is sometimes employed, as shown in the engraving; this prevents the risk of losing the screw.

The spring point in the general construction of hair dividers, if sprung out far, is very weak and unsteady; therefore it should be used to adjust as small distances as possible. For copying drawings or transferring distances, the hair spring can scarcely be required if the head joint move properly. It is only for dividing distances into equal parts that it is found practically convenient. -

The above described ordinary form of hair dividers,

besides being defective in the extreme weakness of the point, when it is sprung out some distance, has also another defect—that of the spring occupying a position up the interior of the dividers, which prevents the hollows in the sides being sufficiently deep to enable the instrument to be easily opened. This is very objectionable, as compasses should have the angle of the hollows in the sides as acute as possible, that they may open readily by merely pressing the thumb and finger on the opposite sides.

SECTION.

Improved Hair Spring.

The construction of hair dividers which the author has found the best, is to place the adjustment much nearer the point than is usual, and to attach the spring entirely below the hollows. In this manner the point may be made very firm, and the hollows as deep as required. The principle of this construction is shown in the illustration above. The author has also applied this point to compasses and bows; however, with other instruments its merits are not equivalent to its expense.

For dividing small distances, the spring bow dividers are used, which are described at page 43.

For the sake of rendering dividers portable, a sheath may be fitted over the points; they are then called sheath dividers. This sheath may be rendered very useful for the purpose for which sheath dividers are often required, that is, for taking off dimensions from working drawings, if the sheath is fitted as is

shown in the illustration, up to the head joint; the
ordinary manner is to make it go merely over the

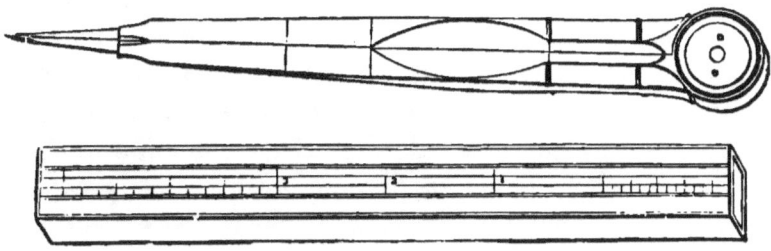

Improved Sheath Dividers.

points. In this new way the sheath is made much
longer, a convenience which allows the leading useful
scales to be cut upon it.

Pillar Dividers, closed.   Napier Dividers, closed.

There are two other forms of pocket dividers, one
called Pillar dividers, the other Napier dividers. These
are constructed with a knuckle joint upon each leg at
about the centre, which enables the point to turn in
to render the compasses portable; they are the most
convenient form of pocket dividers. See also Napier
compasses, at page 47.

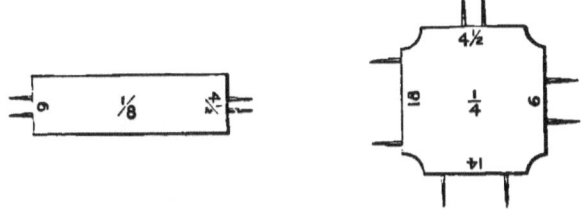

Brick Gauges.

In architectural drawings, the thickness of brick walls

is a quantity constantly recurring. For this purpose the author has made a very cheap divider, which may be used for eighth- or quarter-inch scales with advantage. It consists of a plain piece of metal with two fixed points at each end, representing respectively four-and-a-half and nine-inch work. It will prick off wall thicknesses more exactly than can be done from the edge of the scale, which is liable to become worn. For greater thickness than nine inches, the spaces are set off as many times as required. These brick gauges are also made with four pairs of points on the four sides of a small square, to set off $4\frac{1}{2}$, 9, 14, and 18-inch work.

There are several other forms of dividers not of sufficient importance to require details, as those above described embrace them all in principle.

CHAPTER V.

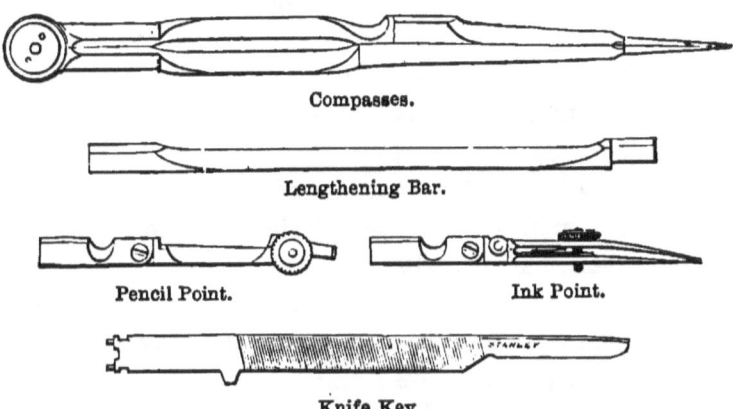

Compasses.

Lengthening Bar.

Pencil Point.   Ink Point.

Knife Key.

DRAWING COMPASSES, in technical terms, are understood
to be compasses which have one or both of the points
or legs moveable. When the moveable point is taken
away, a socket or fitting is left, upon which a point to
carry ink or pencil may be placed, so as to produce a
distinct circle on paper.

The plain dividers, which are not used for producing
circles, but for dividing and measuring only, have been
already described; the same description will answer for
compasses of the most simple kind in all particulars,
except in that of the moveable leg and the additional
points. The points need but little description further
than that conveyed by the illustration above. The

moveable plain point, when it is placed in the compasses, appears exactly like the fixed leg, and forms with
it a simple pair of dividers. 'The ink point resembles
the lifting-nib drawing pen already described, with
the addition of a joint above it, which turns down to
render the pen vertical when it is placed in the compasses.   The pencil point is a holder for a pencil, and
consists of a split tube, in which the pencil is secured by
means of a clamping screw ; it is jointed above similarly
to the ink point.   In addition to the ink and pencil
points, compasses have generally a lengthening bar,
which is used to extend the leg of the compasses, so as
to enable them to strike larger circles.   The lengthening
bar is a plain bar of metal, so constructed that it may
be fixed by one end to the compasses, after the plain
point is removed, and that the other end may carry the
ink or pencil point.   Compasses have occasionally two
lengthening bars, which fit into each other as well as
into the compasses and points.

A knife key is generally supplied with what is technically called the *half-set*,—that is, compasses, points,
and lengthening bar.   It consists of a knife for cutting
pencils, generally a bad one ; a file for sharpening the
lead ; and a key to fit the head-joint of the compasses
—the last being the only useful part about it.   The
knife key would be often better left out of sets of instruments given to young beginners, as it is a frequent
source of amusement to spoil the working of the joint
of the compasses, by tightening and loosening it continually, after it has been properly adjusted by the
maker.

There are two methods of fitting the points into compasses ; one of these, which cannot be recommended,

is used for common instruments. By this method the point is fitted into an angular hole, and fixed by a screw; this screw is very objectionable, as it will soon slip the

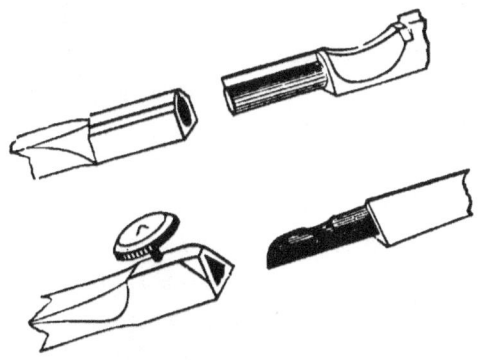

thread, and is liable to be lost; neither does it hold the point perfectly; however, as it is much the easier fitting to make, it has some value for cheap instruments. The general construction of this fitting is shown in the lower illustration.

The *slip-joint* is the best kind of compass fitting. It consists of a round hole or socket, slit up on one side to the entire depth, and a corresponding fitting or *shank*, which is a round pin with a web along one side to fit into the slit. The socket by being slit forms a kind of spring round the shank, which allows for the wear occasioned by the continual changing of the points; and if properly made, maintains at all times a sound fitting joint. It is clearly shown in the upper illustration.

The ink point, pencil point, and lengthening bar form the fittings for every description of compasses, and are seldom varied in any particular. The compasses are differently constructed in the joints and the fixed points. Those described which resemble dividers are

termed single-jointed compasses ; these are the cheapest
and least efficient kind, but answer pretty well for ar-

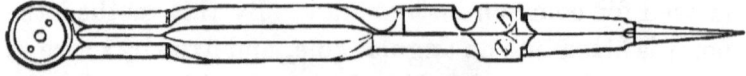

Double Joint Compasses.

chitectural drawing, in which a large circle is seldom
required.   For mechanical drawing, where the circle is
constant, double-jointed compasses are imperative ; the
distinguishing feature of this kind of compasses is a
joint at about the middle of each leg, which will turn
down so as to bring the point vertical with the drawing
when the compasses are opened. It is this that renders
the double joints more accurate than the single ones, as
a vertical point will hold upon the surface of a drawing
more steadily and exactly to position than an oblique
one, besides which it will puncture the paper less.

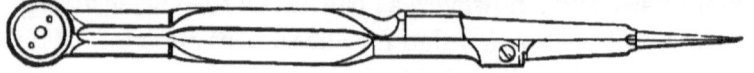

Turn-down Joint Compasses.

Compasses are sometimes made with one turn-down
joint to the fixed point only: these are nearly as useful
as the double-jointed ones, as they form double-jointed
compasses when they are used with ink or pencil point ;
each of these points having, as already described, a joint
in itself.

The plain points of compasses are either of pointed
steel, or are constructed to hold needles, which form the
points.   The plain points are sometimes left triangular
to the extreme, but they are more generally rounded,
or what is technically termed *needled*.   It is a very
common fault in rounding the points of compasses, that

they are left much too weak, and are, therefore, unsteady to use. This is done to give the point the appearance of fineness, which is in itself objectionable; the point should be very sharp, but strong; if the plain point compasses are properly set, they will puncture the paper less than when a needle point is used.

The writer has found the best form of compass point, or that which will puncture the paper least, and keep sharp the longest, to be a rather obtuse cone, of which a very much enlarged sketch is given in the margin. As the cone is required to extend a very short distance up the point, it need be of no obstruction to the sight.

Points to carry needles, as illustrated below, are made generally by slitting up the point of the compasses, so as to form a pair of jaws, in the inside of each

Needle Point.

of which a narrow groove is made close to the edge. The needle is placed in the grooves, and is held by a screw, which clamps the jaws together. This point is the most popular for the best class of compasses; certainly it has one great merit—that a point of perfect sharpness may at any time be provided by a common household needle. Another convenience is, that a shoulder is formed above the needle, which prevents its entering too far into the paper. The common objections to this point are, that if the needle does not fit the groove perfectly, it is liable to be shaky; and that the screw which clamps the needle, although made awkwardly small to

D

use, is yet a great impediment to the easy observation of the point.

The Swiss and French make the needles which they place in compasses with a shoulder near to the extreme point; this answers pretty well, but their manner of fixing the needle, by placing it up a tubular point and jamming a screw against it, is very bad : as the screw has but a few threads to hold by, it is soon spoiled. The manner of holding needles for compasses by a point similar to the ever-pointed pencil, is also very defective.

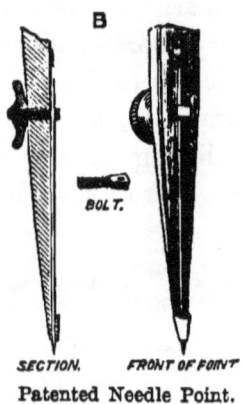

Patented Needle Point.

In the point above illustrated, which the author has patented, and which has been already described as applied to the pricker, the needle is held by passing it through the head of a bolt, and drawing it tightly to the leg of the compasses, by a milled-edged nut. This has many advantages; it does not impede the vision, as the needle is clamped a considerable distance from the point of the compasses, and in such a manner that it cannot slip. The needle is also held quite firmly at the point by being pressed into a conical hole, whose apex is the point of the compasses; this hole has no slit, therefore the end of the compasses does not drill the paper; another feature is, that the needle is easily replaced.

The author has also patented another manner of holding needles, by which, when any pressure is placed

Enlarged Section of Patented Spring Point.

upon the needle, it springs up the point of the compasses, and prevents them making a hole beyond the surface of the paper. This is a convenient point for a heavy hand, or for drawing tracings, but perhaps not to be recommended so much as the last described for ordinary work, principally from the reason that it is difficult to find the centre from which a circle has been struck. Both the above points may be supplied to one pair of compasses, and will by changing be adapted to all requirements for most refined work.

Horn Centre.

When many circles are to be struck from one centre, or many minute arcs are required to be drawn, as in tops and bottoms of cog-wheels, when the last described form of point is not used, it is usual to place a small piece of transparent horn over the centre from which the circles are required to be struck. Small discs of horn, called *Horn Centres*, are the best for the purpose; they are made about the size of the illustration, and have three minute steel points protruding from the under side, to prevent the centre shifting when the

compasses are upon it.   After describing the circles in
pencil, it is better not to remove the horn centres until
the drawing is inked in; nevertheless, if it should
be in the way, it is easily replaced.   A few of these
small inexpensive articles will prove very useful to
every mechanical draughtsman who may aspire to neat
drawing with the ordinary compasses.

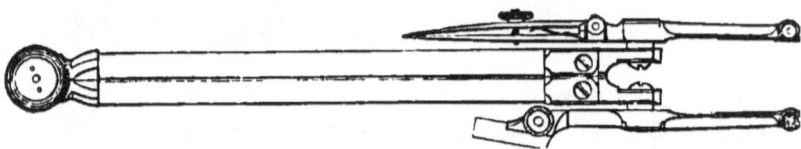

Tubular Compasses, Closed.

The TUBULAR COMPASSES are sometimes used in place
of the compasses and points, for the purposes of which
they answer in every particular.   Being preferred by a
few draughtsmen, they are occasionally packed in complete cases, instead of the compasses and points.   The
advantage claimed for them is that they are complete
in one piece, and do not require any loose points.   It
may be here remarked, particularly if we except the
tubular compasses and pocket compasses, that universal
instruments are generally to be avoided; they profess
to answer many purposes, but really seldom answer any
perfectly, neither is there economy in their use, as they
cost as much as separate instruments.

In construction, the tubular compasses have points
as the other compasses already described.   The legs are
formed out of a pair of tubes, each of which encloses an
inner tube or bar, fitted that it may slide out in the
manner of the tubes of a telescope, when the compasses
are required to be extended to produce a large circle.
Upon the ends of the inner tubes or bars the points are

jointed, so as to turn down in use to an erect position, in the manner of the compass points described. Besides the turn-down joint, each point has a swivel joint, which allows either the ink or pencil, or plain point, to be turned outwards, so that one only of them may form a point to the compasses. The common defect of this kind of compasses is unsteadiness, owing to the weakness of its construction, which is caused principally by the inner tube having a slot down one side to admit of the introduction of a clamping screw to a nut within the inner tube, for holding the instrument in position.

Tubular compasses are much better if made with a solid bar instead of an inner tube. If a slot be made down the bar, and a corresponding slide fitted into it, and connected with the outer tube, the friction of the slides will be quite sufficient to hold the point steadily. In any position, dispensing with the very objectionable clamping screw.

Tubular compasses are very frequently badly made; in selecting one, the tubes should be pulled out as far as they will go, the compasses opened into a straight line, and the points turned down as if for producing a large circle; by taking one point in each hand and twisting the points with a rocking motion, it may be easily ascertained if the work is sound. In other points it is only necessary to observe that the joints move evenly and the tubes firmly.

CHAPTER VI.

INSTRUMENTS FOR PRODUCING SMALL CIRCLES— BOWS—SPRING BOWS.

FOR drawing small circles, a kind of compasses is used termed *Bows*. These are distinguished from very small ordinary compasses by the addition of a handle above the head joint, which is made of a suitable size to roll conveniently between the points of the thumb and forefinger. The handle is fitted so that it partially encloses the head, and forms a part of the joint. The points of bows are not made changeable, in the manner of points of compasses, but each instrument is constructed for one purpose only—that is, for producing small circles, either in ink or in pencil. Therefore, in a set of instruments there are two pairs of bows; these are called respectively ink bows and pencil bows.

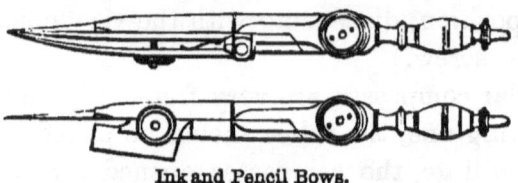

Ink and Pencil Bows.

The above illustration represents the usual form of single-jointed bows sufficiently clear to need no further description. This kind, although in extensive use, are not of very scientific construction; the joints are unnecessarily close to the points, which throws the ink point, when the bow is much opened, very much out of a vertical position, and makes it necessary for the inner nib of the ink point to scratch into the paper before the outer nib touches, to produce the circle.

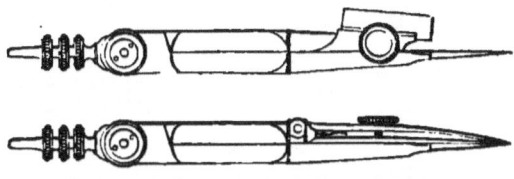

Improved Form of Single-jointed Bows.

Single-jointed bows are very much improved by making the joint nearer to the top and the handle shorter, as is shown in the illustration above. The handle is also much better if milled with several mills —being rough, it is held with less risk of slipping out of the fingers. Another improvement is to make the angles of the sides of the bows slope inwards instead of outwards, which enables them to be opened more readily with the point of the thumb. At best, single-jointed bows are very imperfect, from the fact that the points of all circle-producing instruments should be vertical to the paper when in use; they are nevertheless sometimes preferred by architects, who do not require the circle very frequently. The management of a single-jointed instrument is also somewhat less tedious than a double-jointed one, as one joint only is required to be moved to adjust it, instead of three. It must, nevertheless, be borne in mind that the double-jointed instrument produces the most perfect work.

In principle, double-jointed bows exactly resemble double-jointed compasses, except that they have a handle above the head-joint, and that the points do not change, as has been already observed of single-jointed bows. The old form of double-jointed bows, as represented below, are those in very general use. They may be much improved in form and construc-

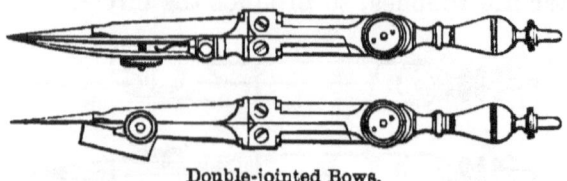

Double-jointed Bows.

tion, in a similar manner to that described for the single-jointed bows, as in these also full advantage is not taken of the size of the instrument to produce a proportionate circle. Bows of the above construction, of three inches in length, when the points are vertical, only produce a circle of three inches diameter. This is particularly caused by the shortness of the middle part of the bows between the head and knuckle joints.

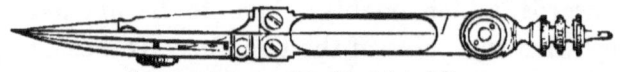

Improved form of Double-jointed Bows.

Bows combining the improvements suggested, are illustrated in the engraving above. By lengthening the central part, or *body*, as it is termed, the bow will produce a circle of five inches instead of three inches diameter, and this without lengthening the entire instrument. It is important, to prevent the necessity of constantly changing the instrument for another, that it should produce as large a circle as possible consistent with its length, this being the circle-producing instrument most constantly in the hands of the draughtsman.

The plain points of double-jointed bows are often constructed to carry needles, and except being smaller, resemble in every respect the points of compasses of the class already described ; they are also occasionally made with a spring point, as shown in the second illustration below.

Bow-point to hold Needles.

Spring Bow-point.

The author's patented point, described at page 20, is especially adapted to bows, from the little obstruction which it offers to observation of the points. For the

French Needle-point.

ink bows of this make, instead of using the common needle, the French needle is used ; this is shown in the illustration above, and consists of a piece of fine wire upon which a shoulder and point are formed by filing away the end, so as to leave the point at one edge of the circumference. The value of this kind of needle for the ink bow is that it is easily adjusted to the setting of the pen.

Bows are sometimes made with exchangeable points, as described for compasses, or the points are made to turn round, as in the tubular compasses; both ways are very objectionable. In the first kind, the points being short, in taking them out, the ink soils the fingers, and in the swivel kind the ink runs over the instrument. The saving in cost is very trifling. They are called universal bows—and should be universally avoided.

The author has devoted considerable time—not, he must say, with success, in endeavouring to remedy the awkwardness which is always felt in the simultaneous management of the three joints of double-jointed bows. His aim has been to make the points keep an erect position, without the trouble and loss of time in setting each point separately.

Perhaps the best method is to have a bar fixed below one of the joints of the leg of an ordinary double-jointed bow, and a tube fixed to the other leg, in opposite position, each at right angles with the point. The bar sliding in the tube holds the points in a vertical position at all openings of the bow. In this the form of instrument is not materially altered, so that a draughtsman may have the convenience of his points being self-acting, without the necessity of acquiring the

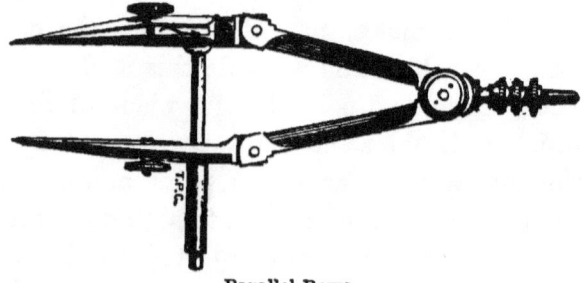

Parallel Bows.

peculiar management of a new form of instrument. This plan, however, is not satisfactory, and is only given as the writer's best attempt at producing something which would be a real convenience, if it could be effectually managed.

SPRING Bows are small compasses in which the head joint is entirely dispensed with, the sides of the bows being of tempered steel, and forming two springs, which cause the points to diverge, a screwed bolt is jointed to one of the spring sides and passed through an opposite hole in the other, a milled-edged nut upon the bolt sets the bows to the required radius. Three spring bows form a set—namely, dividers, ink bows, and pencil bows; they are very useful instruments, and are put in all full-set cases.

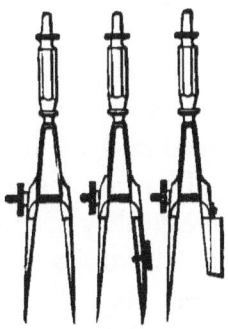

Set of Spring Bows.

The dividers are mostly used for setting off distances, as the thickness of walls, etc. The ink and pencil bows of the ordinary size are adapted to draw circles from the fortieth of an inch to an inch and a quarter diameter.

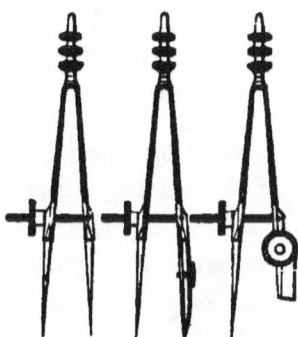

Long Form of Spring Bows.

The ink bows do not draw over three-quarters of an inch diameter very perfectly, as the point becomes too oblique, or technically *kicks*. To obviate this, the body or spring part of the bows may be made as long as is consistent with leaving sufficient handle to turn easily with the fingers. This is an improvement for the larger circles; however with length they lose steadiness, and altogether are not perhaps to be preferred to the previous ones.

The pattern shown above, which the author employs, will draw circles from three-quarters of an inch upwards better than can be produced by spring bows of the ordinary form. Spring bows are frequently made with the points to carry needles; they are more expensive, and practically not so good, from the reason that the points, which should be clearly seen, are much obstructed by the jaws which hold the needles. The points also require very nice workmanship, or the needles will be weak and shaky.

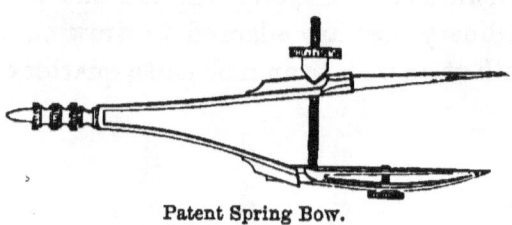

Patent Spring Bow.

In the above-described spring bows, the points take the oblique angle of the spring, which is only of consequence in the ink bow. By the plan that is shown above, and which has been already described for double-joint bows, the points may be held erect. For this it is found practically the best to place the knuckle joint to the ink bow, only at a very short distance above where the bolt enters the ink-point side, and to screw the bolt firmly into this side, instead of making it loose in the ordinary manner. By this method, the ink point is held erect in all positions, and will draw a clear circle to the full radius; but it is perhaps rather less steady

for very small circles, a fault that almost counter-
balances the merit of its vertical position.

There is another kind of spring bow, of Swiss invention, which is very useful for some purposes, particularly for lithography—it is called the Pump Bow.   In
this bow, one point only springs ; the other is a tube
with a rod passing through it, one end of which forms

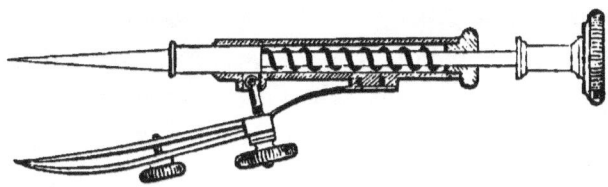

Pump Bow.

the true point or centre of the instrument, the other
end of the rod forms a handle. The spring which
carries the ink point is fastened to the tube or jacket,
and adjusted with a bolt, as in the ordinary spring
bows. Within the tube is a spiral spring, which lifts
the ink point off the drawing. To use this bow, the
top of the rod is held with one hand whilst the milled
head just below, which is connected with the tube, is
pressed down with the other hand until the pen touches.
The milled head is then rolled between the thumb and
finger the circumference required. It will be observed
that the point does not turn, being held without
motion by the top, therefore it does not make a perceptible centre mark. The fault of these bows is their
clumsiness—they require the use of both hands to
make a small circle.

CHAPTER VII.

PORTABLE COMPASSES—NAPIER COMPASSES—PILLAR COMPASSES, ETC.

BESIDES the compasses already described, there is another class made expressly to be carried in the pocket; they are very useful to the architect and engineer, in addition to the ordinary case of instruments, being frequently of especial convenience to apply to the drawings when works are being carried out, either for the sketching of details incomprehensible to the workman, or for reference to dimensions according to the scale.

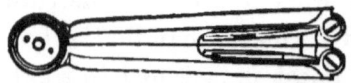

Napier Compasses, closed.

The NAPIER COMPASSES are, perhaps, the best and most popular kind of pocket compasses, their especial merit being their compact form. The usual dimensions when closed, as shown in the figure above, are three inches long, three-quarters of an inch wide, and less than half an inch in thickness. When they are opened they form a five-inch instrument. The portability of the Napier compasses is obtained by the sides being hollowed out from the inside, so as to admit of the points folding into them. This form also gives the sides great strength and lightness, and protects the points from the intrusion of small articles which may

be in the pocket. The knuckle joints, which allow the
points to fold in, also act as the knee joints of double-
jointed compasses, and permit the points to be kept
erect when in use. The parts which fold from the
knuckle joints, technically called straps, have swivel
joints at their ends, which carry reversible points, the
one for ink point and plain point; the other for pencil
point and plain point. Either of these reversible points
will turn outwards, so that the compasses may be used
as dividers, pencil, or ink compasses.

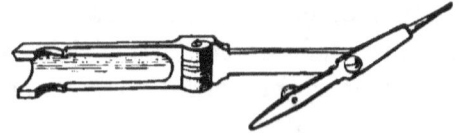

One leg of Napier Compasses, open.

The cut, which represents part of one leg of the
Napier compasses as seen from the inside when opened,
will sufficiently show the construction and proportions.

The following qualities are expected in well-made
Napier compasses. When they are closed, the sides
should meet perfectly, as should also the backs of the
straps; the inside hollows should be cleanly finished,
and as large as the sides will admit of, particularly
that sufficient space may be provided for a useful ink
point; the head joint should move smoothly, and
the knee and swivel joints should move stiffly. They
are a difficult instrument to make perfectly.

Napier compasses are sometimes packed in a pocket
case, frequently with a small ivory scale, or with a
set of small scales; in this manner they are very
useful.

Pocket compasses as illustrated below were very popular in this country a few years since. They were principally of Swiss and French manufacture; the

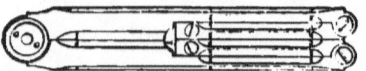

Swiss Pattern Pocket Compasses, closed.

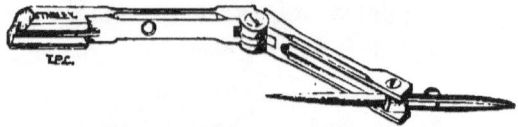

Ditto, part of one leg shown open.

form is rather more elegant, but in most respects similar to our Napier compasses, excepting that the swivel joints are made on the same plane as the head joint, instead of at right angles to it. In this construction, to obtain sufficient strength, two straps are required to each swivel, between which the point revolves. The lower illustration of one side of the pair of these compasses shows the principle of construction. The upper shows the instrument when closed.

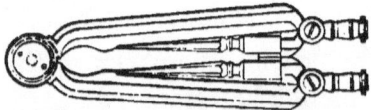 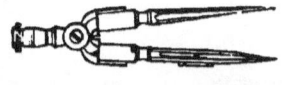

Pillar Compasses, closed.　　One leg of Pillar Compasses drawn out.

PILLAR COMPASSES, although less portable than the Napier, being somewhat wider, have some merits, which make them preferred by many, as they answer more completely the uses of a set of drawing instruments, and may be recommended to young beginners who cannot afford to purchase a set of instruments of good quality. They will answer all purposes much better than an inferior set of instruments, which will cost as much as a very respectably made pair of pillar compasses.

The foregoing illustration represents the instrument when closed, which will sufficiently indicate its general form. It will be observed that the sides are bowed away from the head joint so as to allow the points to fold between them. Each side is formed of a tube, which admits either point to slide up it, and to be fixed by a clip, so that the points may be reversed; for instance —if the ink point is required to be used, the plain point is pressed up the side; if the plain point is required, the ink point is pressed up, leaving the plain point in view. Thus by reversing the points, they will answer the purposes of the Napier compasses already described, either for dividers, pencil or ink compasses. Further, as the points pull entirely out of the sides, they may be used as separate instruments, the one answering the purposes of an ink bow, and the other that of a pencil bow, to be used for producing small circles. The ink point or bow is illustrated below.

Bar Pillar Compasses, showing one point and bar pulled out.

Pillar compasses are frequently made with lengthening bars, which are tubular pieces, similar to the sides of the compasses; they fit between the compasses and points, and are used to extend them for producing

E

larger circles; they also form sheaths to protect the points when the instrument is out of use.

In selecting a pair of pillar compasses, it should be observed that the points, when reversed, fit without any shake; that they will also meet in any position without crossing; and that the ink point, when pulled out, will close so as to produce a very small circle.

Some little variation is occasionally made in pillar compasses; as, for instance, the omission of the handles to the points. The points of the various kinds of pocket compasses are also sometimes made to carry needles, or have adjusting springs similar to the hair dividers, neither of which schemes, although more expensive, is so useful as the plain points. There are some other kinds of pocket compasses, of which a description is not necessary, as they are very little used, and in many respects inferior; such as the breeches compasses, turn-in-tube compasses, etc.

CHAPTER VIII.

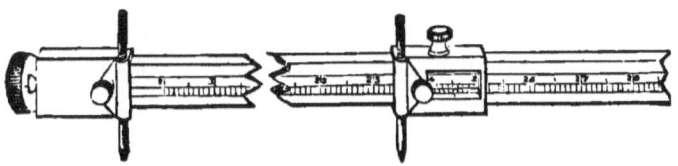

Divided Beam Compasses (Ordnance Pattern).

BEAM COMPASSES have at all times been considered the most perfect instrument for transferring exact measurements or for setting off distances which exceed a few inches. For drawing purposes, however, their use is mostly confined to radii or measurements of from fifteen to seventy inches.

The most simple construction of beam compasses consists of an angular bar of wood, which is generally well-seasoned mahogany. Upon this, two instruments termed beam heads are fitted, in such a manner that the bar may slide easily through them, the general construction of which may be seen from the illustration. A clamping screw attached to one side of the beam head will fix it in any part of its course along the beam. Upon each head a socket is constructed to carry a plain point, exchangeable for an ink or a pencil point. For exact purposes the beam head placed at the end of the beam has a fine adjustment, which moves the point a short distance, to correct any error in the first rough setting of the instrument. This

adjustment generally consists of a milled-head screw,
which passes through a nut fixed upon the end of the
beam; the screw has a shouldered fitting to the beam
head, which it carries with its motion.   For scientific
purposes a divided cylinder is attached to the milled
head, to indicate the quantity the screw may be turned,
and consequently the distance that the point will be
moved, according with the known number of the
threads of the screw per inch : this is called a micro-
meter adjustment.

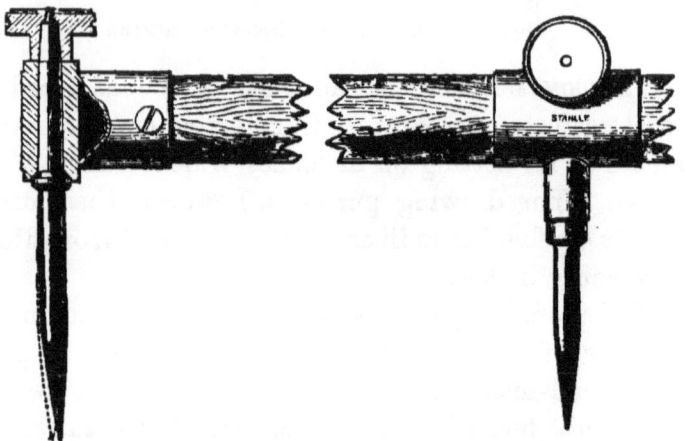

Colonel Strange's Adjusting Beam Compasses.

The author has made some beam compasses for the
Government public works in India, the adjustment of
which is the invention of the late Colonel Strange,
Inspector of Scientific Instruments for India.   This
adjustment is extremely simple and very effective; it is
shown in section in the illustration above.   It consists
of a long pin, shouldered at about the centre, the
upper portion from the shoulder is accurately fitted to
the beam head, so that it will turn with even motion.

A milled head placed on the top of the pin secures it
in its position, and forms the means by which the pin
may be revolved. The point of the pin below the
·shoulder is bent, so that the extreme point is about the
tenth of an inch eccentric to its axis. It will be readily
seen that by moving the milled head any portion of a
revolution, the point will either recede from, or advance towards, the point upon the second beam head.
The merit of this adjustment is, that the hand is
directly connected with the point, so that you may
feel as well as observe the exact movement upon the
point; it is therefore very expeditious.

Beam compasses are occasionally divided to a scale
upon a slip of holly or boxwood, which is inlaid on
one side of the beam, the scale being read through
openings cut on one side of the heads, which are
bevelled down and divided with a *vernier reading* (for
description of vernier, see protractors further on), as
is shown in the illustration at the commencement of
the chapter.

Divided beams are very little used at the present
time; they have a fault which renders them, although
costly, less exact for setting off distances from the ordinary drawing scale. This arises from the beam requiring to be made of slight wood, for lightness, which
makes it very liable to warp and to throw the points
out of their true vertical position, besides shrinkage,
which all wood is subject to, thus rendering the readings upon the beam incorrect. There is also in setting
the heads a kind of torsion, difficult to overcome, so
that if they be set twice to the same distance, the
readings will seldom exactly correspond. In using
the beam compasses for setting off distances to scale

in surveying, it is better to have a separate scale, which will answer the purpose of a straight-edge; this will be referred to further on. Divisions of inches and parts along the beam are very useful for mechanical drawing, in which instance the beam is used merely to produce curves, and positive accuracy is not required.

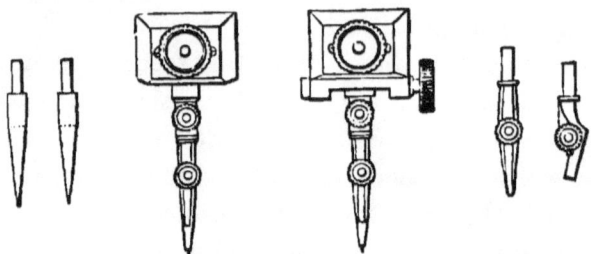

Portable or Swiss Pattern Beam Heads.

The Swiss manner of making the heads of beam compasses is to leave the top side of the head open, so that instead of the beam being required to pass through the heads, these heads fix on the edge of any lath or straight-edge, the clamping screw being placed on the side of the head instead of at the top. To prevent the clamping screw making an impression on the straight-edge, it is made to carry a plate before it guided by two steady pins. These beam compasses are sometimes made with an adjustment, as is shown in the illustration ; they are also made to carry ink and pencil points.

The Swiss pattern beam compasses are now made in this country, and have in a great measure superseded all others, from their cheap, portable, and effective construction, with the peculiar merit that any rule, rod, or straight-edge will make a beam, in this way enabling the draughtsman to have beams of different lengths, by making use of the straight-edges, etc., in general office use. They are in every way sufficient for the architect or the mechanical engineer.

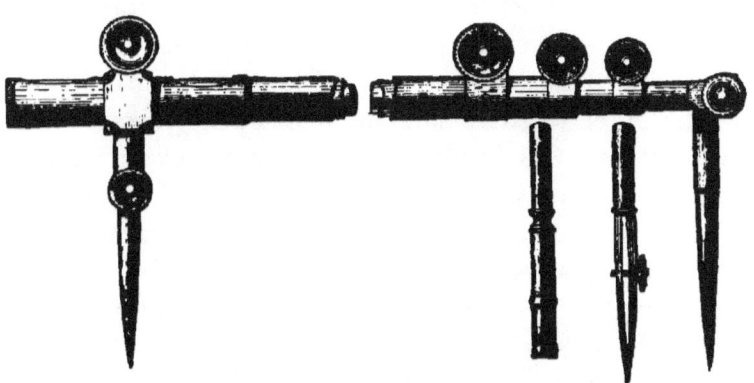

Tubular-Beam Compasses.

The TUBULAR-BEAM COMPASSES, as illustrated, are of a different principle of construction from those already described. The beam, instead of being solid, consists of a series of tubes fitted to slide one within the other, in the manner of the tubes of a telescope. When the tubes are extended, they form a beam of from 24 to 30 inches in length. By a clamping screw at the end of each tube the beams are fixed at the required extension. One of the beam heads slides along the outer or larger tube, as in the ordinary beam compasses. The other beam head is constructed upon the end of the inner .or smaller tube, being fixed, and not sliding in the ordinary manner. The fixed head is merely a clamping socket to carry pencil or ink point.

When an adjustment is connected with the tubular-beam compasses, a kind of head is formed upon the inner tube, through which the adjusting screw passes.

The principle of the adjustment is generally that of drawing the head inwardly by a milled-head screw, which acts against the outward pressure of a spiral spring placed within the tube.

The tubular-beam compasses are very portable and complete in themselves, but are very unsteady and awkward to use; one tube is liable to slide entirely out of the other when setting the instrument to nearly its full extent; the heads also turn round at all angles, requiring much adjustment before describing a circle. These beams may answer very well for the architect's use, who only very occasionally requires a large circle or curve, but are not at all adapted for setting off distances or for dividing.

There are many other methods of making beam compasses and appliances connected with them, the description of which would occupy space and be of little use to the draughtsman or mechanic; as, for instance, portable wooden and tubular beams jointed together in various ways; there are also many peculiarities of adjustment.

A few remarks may, however, be made on some of the most common appliances; one of these is a fixed centre, which is sometimes supplied for the purpose of preventing the point, used as a centre in striking circles, from making a hole in the drawing. This centre consists of a square piece of metal, with a pin projecting from the upper side, which fits into the socket of the point of the beam head, so as to form a pivot for the beam to turn upon. Its use will not be found worth its expense, as it covers up the centre which is required to be seen; further, a plain horn centre will answer the purpose practically better. Castors, which are sometimes added, are also very unnecessary.

A few observations may be made on the manner of using the beam compasses, whether for striking a circle or setting off a distance.—The beam should not be touched, if possible, but should be held by the heads only. It should be held as nearly as possible with the points vertical to the paper. In striking a circle, neither the head which carries the centre nor the beam should be touched; after the point is set in its place, the hand should move the head steadily which carries the pencil or ink point, leaving the other head free. It will not pierce a considerable hole, if it has a conical point, as shown at page 33, and the weight of the beam will be sufficient to keep it to its centre. Points which carry needles are the best for striking circles; plain points are the best for setting off distances. Good beam compasses are fitted with both kinds, which are made exchangeable.

For setting off exact quantities by the beam compasses, a standard rod should be employed. One of the best plans of making this is to take a rod of thoroughly seasoned deal about an inch and a half square; it should be cut out a year or so before it is used: it will then be found to change very little in length by either temperature or humidity. Insert into this at every foot a piece of brass sufficiently large to take one line across the rod. At the end of the rod insert and screw down a plate of brass the width of the rod, and thirteen inches long. This should be divided by engine twelve inches on the one edge into one thousand parts, and on the other into inches, each inch subdivided into one hundred parts. From these divisions all scales are to be calculated. A variation will sometimes be found convenient in this division by having one edge metre where such is required instead of the decimal foot. The length of the rod will vary according to requirements; the ordinary lengths are ten, five, and three feet. If the standard is placed or fixed in the office, all important dimensions should be taken from it by the beam compasses, details being afterwards filled in by the ordinary scale. The writer has also supplied these standards for use in large engineers' shops, with appropriate beam compasses. The standard being fixed, shop rules have been made that all important dimensions should be referred to this by the beam compasses. He is informed that the plan is highly satisfactory, obviating much of the inaccurate two-feet-rule measurements. It will be easily understood that the standard need not be of the entire length required, as lengths on metal are easily continued by scratches on the metal by the beam point and centre-punch marks.

CHAPTER IX.

INSTRUMENTS FOR STRIKING ARCS OR CIRCLES OF HIGH
RADII—HOOKE'S INSTRUMENT—CENTROGRAPH, ETC.

THERE is not, perhaps, any portion of practical geometry that has had a larger amount of thought and
experiment devoted to it by our mathematicians than
the production of arcs of high radii by means of
mathematical instruments. That the subject should
be worthy of great consideration will be easily conceived, if we consider the important philosophical
operations into which arcs of high radii enter, such
as the projection of the sphere, and other constantly
recurring instances, where the beam compasses would
be quite inadequate to produce the required curve.

For professional purposes, such as the curves upon
railway plans, lines for strengthening girders, etc.,
the general method of describing arcs of high radii has
been by means of templets, termed technically radius
curves or railway curves; these will be described with
other curves in a future chapter. The direct instru-
mental means by which arcs of high radii may be
constructed is taken up here, somewhat exceptionally
to the general practical character of this work, to
make the subject as complete as possible for giving
the best means of producing all geometrical forms,
although the *best* means known may not be perfect
for general practice.

If we examine the many recorded experiments and
instruments which have been constructed for the pro-

duction of arcs of high radii, we shall find that there
are only two principles of construction which may be
considered satisfactory in the results they produce.   In
one of these, the line of curvature is produced by an
oblique rolling surface ; in the other, the curvature
is derived from the motion of an angle between two
points.

The best instrument for producing arcs of high radii
by oblique rolling contact was invented by Dr. Hooke,
and afterwards somewhat improved by Mr. George
Adams, an eminent mathematical instrument maker of
the last century.

In construction, this instrument (described and illustrated in Adams' " Geometrical and Graphical Essays ")
consists of a kind of triangular carriage, supported
upon three milled-edge wheels.   The axis of one of
these wheels is fixed in a line direct with the marking
point, placed upon the opposite centre of one side of
the triangle.   The axes of the other two wheels are
moveable upon centres, so that they may be set either
perfectly parallel with the axis of the first wheel, or
that the lines of their axes converge, and meet the line
of the axis of the first wheel in any required distance.
The two moveable wheels have each a protracted scale,
by means of which any required degree of convergence
may be exactly indicated.   The rule of convergence,
or angle of the moveable wheels required to produce a
given arc, is ascertained by a printed table supplied
with the instrument.

To describe an arc after the instrument is set, it is
rolled along the surface of the drawing, the ink or
pencil point tracing a line exactly corresponding with
the track of the instrument, which also corresponds

with the obliquity of the wheels. Thus, if the wheels be set so that their axes are parallel, a straight line will be produced; if they slightly converge, an arc of very high radius; and so on, according to the rates given in the table.

The foregoing may be considered an experimental instrument, as it has never come into professional use, although it produces arcs of high radius with surprising accuracy. It has, however, one important imperfection, which is, that there are no means of directing it to produce an arc in any given position—as, for instance between two given points—a constant requirement in geometrical drawing.

Undoubtedly the most accurate instruments for producing arcs of high radii have been constructed upon the principle of the motion of an angle between two points, the theory of which is derived from the often-repeated thirty-first proposition of the third book of Euclid's Elements of Geometry: *The angles in the same segment of a circle are equal to one another.* In practice this is well understood by the intelligent mechanic, who, to draw his arc through three given points, fastens together a triangle of slips of wood, so that one of the angles and two of the sides shall touch the given points, through which he draws his arc by the evolution of the angle against two pins which he places in the acute angles.

The author devoted considerable time to experiments based upon this principle, not with great success, in order to arrive at an efficient means of producing arcs which should be so far under the control of the draughtsman that he should be able to produce any required arc in any required position;—the only conditions upon which an instrument of the kind could be of great use, and for the want of which other instruments had failed to be practical.

The instrument illustrated below was constructed by the author, after examination of the qualities and defects of the many instruments that have been made upon the principle described in the last paragraph, some of which are perfect for the production of the arcs, but fail in being insufficiently manageable. This has also defects, and it requires a large surface to work upon.

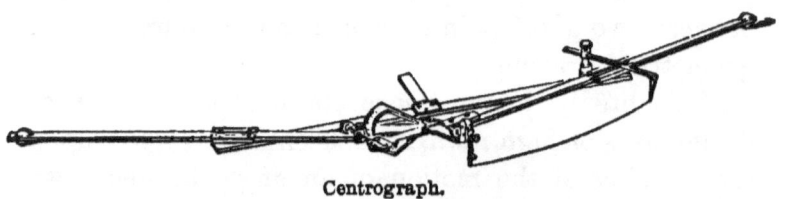

Centrograph.

The CENTROGRAPH, in construction, consists of a moveable *frame*, which forms the angle of curvature, and a stationary *bed*, which supports the foci of evolution. The details of the *frame* are two arms of metal, which are united by a joint, by which they may be inclined at any angle to each other. The joint forms the centre of a protracted or divided segment, which is attached to one of the arms. The other arm is carried over the centre until it meets and reads into an arc divided upon the segment. *A clamp-and-tangent motion** is attached to the segment to enable the

---

* Clamp-and-tangent motion consists of some means of fixing or clamping one part of an instrument in such a manner that it will allow of after adjustment by means of a screw. The screw being generally placed tangent to a divided circle, is called technically a *tangent*.

required angle of the arms to be adjusted with pre-
cision. The arc is divided to half degrees, which read
by the vernier upon the end of the arm to single
minutes. In a line from the centre of the joint an
apparatus to hold ink or pencil points is fixed; this
has an adjustment to it, to permit the point to be
moved a sufficient distance to produce concentric arcs
without moving the bed of the instrument.

The second part of the instrument, the stationary *bed*,
consists of a flat rule of metal with a piece projecting
from about the centre to keep it square and solid.
Near *each* end of this is attached, upon a moveable
centre, a kind of box or head, which fits over one of
the arms of the frame first described, so as to allow
it to slide evenly through, the centre of the box
moving horizontally to follow any direction the arms
may take. Upon the under side of the bed are projecting needle points to prevent it slipping upon the
surface of the drawing.

There is another part of the instrument which is fixed
near one end of the bed, not connected with the action
of the instrument, but used merely as an indicator for
setting it. This indicator consists of a kind of post
which carries a light steel bar, moveable upon the post
in any horizontal direction. At the end of the light
steel bar is a point carried down to the surface upon
which the instrument stands, so as to meet the point
of the pen or pencil when moved with the frame to
that end of the instrument. A clamping screw fixes
the indicator in any position in which it may be placed.

By the reading of the divided arc, the arms of the
instrument may be set to any angle up to thirty degrees,
which is sufficient to produce arcs from thirty inches

to infinite radius, or even a straight line. There is a table affixed to the inside of the lid of the case in which the instrument is packed, giving angles in relation to radii, at as frequent intervals as are generally required, of from 30 to 10,000 inches.

In using the instrument, one hand is placed upon the bed, and the other steadily moves the head which carries the ink or pencil point.

To produce any arc between two given points, the arc of the instrument is first set by the table of radii, the indicator is then adjusted so that it will exactly meet the point of the pen or pencil, whichever is used; the indicator point and the ink or pencil point are then each placed upon one of the points between which the arc is required to be drawn. In this position of the instrument, the bed is lightly pressed upon the drawing, so as to prevent it slipping; if the ink or pencil point be now drawn to the point of the indicator, the required arc will be produced between the given points.

If concentric arcs are required, it is only necessary to move the adjusting screw of the ink or pencil socket head the amount required. It is better when using the ink point not to supply it with ink until the instrument is in its position. The ink point will turn up off the drawing to enable it to be supplied.

The centrograph above described does not produce a perfect arc, inasmuch as the marking point is not exactly in the angle, and it is adjustable; but the error caused thereby is so infinitesimal, that practically it may be ignored, being impossible of detection, except in arcs of low radii, which the instrument is not intended to produce. In a book recently edited by

Mr. Heather, he has been ungracious enough, after appropriating matter and arrangement of this work, of which he has not mentioned the author of his information, to call attention to the above inaccuracy, which some one, it is presumable, has pointed out to him; as his previous book, as far as it related to drawing instruments, did not indicate a very profound knowledge of the subject.

F

CHAPTER X.

INSTRUMENTS FOR STRIKING ELLIPSES—THE ELLIPTIC TRAMMEL—SEMI-ELLIPTIC TRAMMEL, ELLIPTOGRAPH, ETC.

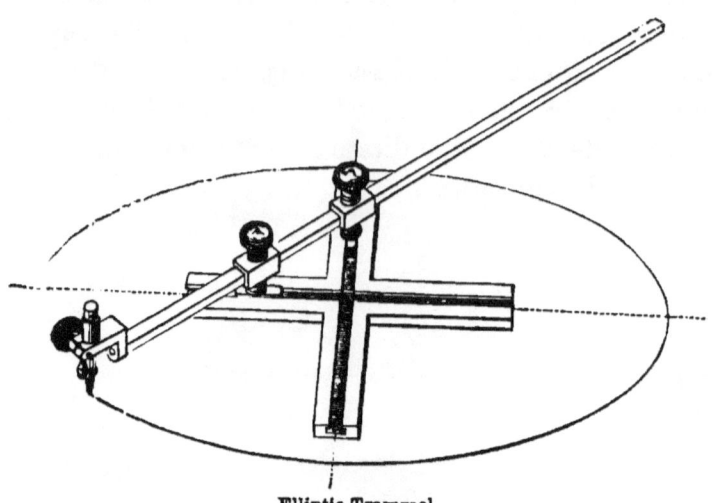

Elliptic Trammel.

THE ELLIPTIC TRAMMEL is the most simple and best known instrument for striking any ellipse which exceeds five inches in its minor axis. The illustration above will convey a correct idea of this instrument as it is generally constructed; the details of which are:—A *bed* of metal in the form of a cross, into which are sunk two similar under-cut grooves at right angles to each other. This cross is held stationary upon the drawing by means of projecting needle points from its under side. Above the cross is placed a bar with two sliding heads similar to the beam compasses described in a former chapter. From the under side of each of these heads a centre is carried down and connected with a sliding slip, which fits into one of the under-cut grooves of the cross. At the end of the bar a socket is fixed, to carry pencil or ink point.

The elliptic trammel is adapted to produce any ellipse whose half length, or major axis, is not greater than the length of the bar, and the difference of whose width, or minor axis, is less than equal to the width of the cross. A small cross would strike an ellipse of any size by having a bar of the required length; but as no greater difference of axes can be produced than the width of the cross, one elliptic trammel is thereby limited to a very short range of ellipses of agreeable and usual form.

The ordinary elliptic trammel used for drawing purposes has a $3\frac{1}{2}$-inch cross. This is useful for drawing an ellipse from six to fifteen inches major axis. The cross is sometimes made much smaller, but it does not work very perfectly, from the necessary shortness of the sliding pieces.

For producing a given ellipse in a given position with the elliptic trammel, it is necessary to draw two lines across each other at right angles; these lines will represent the major and minor axes. Set off half the required major and half the required minor axis from the intersection of their respective lines. Place the cross of the instrument exactly over the lines, observing that the lines drawn from the ends of the grooves on the instrument fall exactly over the axial lines on the drawing. Place the bar or beam in a line with the major axis, which will bring the sliding head which moves in the groove in the minor axis to the centre of the cross. It should then be observed that the sliding head which moves in the major axis groove should be between the other head and the head which carries the pen. Unclamp the screws of both the sliding heads, and slide the bar through them until the pen comes over the mark set off for the required major axis, then clamp the screw of the head which slides in the minor axis groove. This is the setting for the major axis. For the minor axis the pen is moved round, and set upon the mark set off on the minor axis, and in this position the major axis slide is clamped. By moving the pen-head round, the ellipse of required dimensions will be produced in the given position.

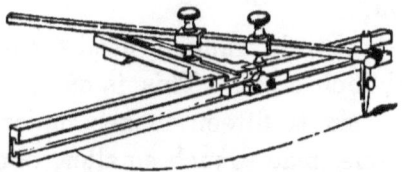

Semi-elliptic Trammel.

The SEMI-ELLIPTIC TRAMMEL, as illustrated above, is found very useful to the civil engineer for the production of elliptic arches, faces of skew-bridges, etc.

Although this instrument will only produce half of an ellipse, it is in all respects for this purpose more perfect than the last described, as no exact limit is made to the size of the semi-ellipse it will produce. It also has the merit of being easily and correctly set, as the face of the instrument may be placed against the line of the major axis, instead of over it. The principle of its action is the same as that of the elliptic trammel, but making use of three arms of the cross only, which enables the slides to pass each other on different planes ; the major axis slide being placed in a vertical position along the front of the instrument, and the minor axis slide, as in the elliptic trammel, on the top. This plan also enables the slides to be made much longer, thereby causing them to work more smoothly. The dispensing with one arm of the cross allows the centre to be approached, so that small semi-ellipses may be produced. As the instrument is generally made, it will produce perfect semi-ellipses of from two to twenty inches major axis.

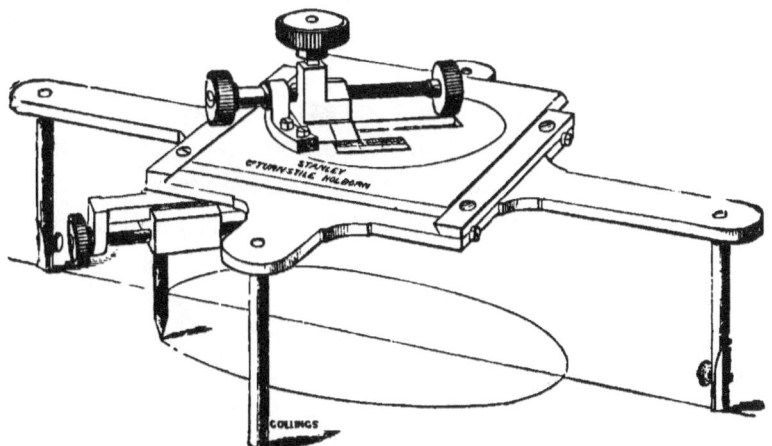

Finney's Elliptograph, with Improvements.

The ELLIPTOGRAPH, of which the following is a description, is the invention of Mr. James Finney, and is known as Finney's elliptograph. The instrument illustrated is different in detail from the original form,* which the author found to be difficult to set or use, as the pen could not be lifted from the paper without moving the instrument, neither were there any means

* The original instrument is described and illustrated in the "Engineer and Machinist's Drawing-book," compiled from the works of M. Le Blanc and MM. Armengand, published by Blackie & Son, 1855.

of setting it to position.   Otherwise the original instrument was correct in geometrical principle, which
is therefore preserved unmodified in the following
description.   If properly constructed, this elliptograph
may be readily applied for the production of ellipses of
from six inches to half an inch major axis, thus taking
the range of small ellipses downwards from the point
where the elliptic trammel ceases to act perfectly.   The
principle of its construction is the combination of a
circular with a rectilinear motion, the details of which
are as follows :—A square piece of flat metal having
four arms projecting from it at right angles, the
centres of which represent the axis of the instrument,
and the whole forming a stage upon which the other
parts are fitted.   The centre of this stage is pierced
by a groove in a line with the major axis, and a slide
fitted into the groove.   Upon the stage is fitted a
square flat piece of metal, nearly the size of the
stage.   This is guided by two dovetailed slips attached
to the stage, so that it will move transversely only.
The centre of this square moveable piece has a disk
fitted into it, as large as it will admit of to leave sufficient strength of metal to retain the square.   The disk
has a groove crossing the centre to nearly the edge of
it, in which a slide is fitted.   The slide has a centre or
axis carried down, which passes through and forms a
second axis in the slide of the stage beneath.   The
slide in the disk has an adjusting screw passing along
one side of it.   Upon the opposite side there is an index
which reads upon a scale engraved upon the disk ; by
this scale the required eccentricity of the ellipse may
be regulated.   Vertically through the centres or axes
of both the slides a square bar is fitted which moves

easily in its fitting. The bar has a milled head at the top end, and a cross piece upon which a scale is divided at the bottom end. Upon the cross piece is an adjustable head fitted to carry pen or pencil point.

The bar being square, and passing through the centre of action, admits of the pen being brought down upon, or lifted off, the drawing when required, without moving the instrument. The cross piece at the bottom of the bar should be in a right line with the groove in the disk above, or the circular motion will not be correct.

At the ends of the arms of the stage are carried down four legs to support the instrument at the proper distance from the drawing. The two legs which are in the major axis have needle points projecting a short distance from one side of their bottom surface. By these the instrument is set, and attached lightly to the drawing, to prevent movement in using.

To set the instrument to produce a given ellipse, it is taken in the hand and turned upside down. In this position the head which carries the pen and pencil is adjusted on its scale to the required *minor* axis. The instrument is then turned upright, and the required *difference* of axis is set off upon the scale of the disk, by adjusting the upper slide to the quantity. For instance, if the ellipse is required to be two inches minor and three inches major axis, the difference being one inch, the head on the lower scale being set to the two inches, the upper scale on the disk is set to the one inch engraved on it, making together three inches, the required major axis.

To set the elliptograph to a given position, an extended major axis is drawn, and a mark is made on the line at three inches distance from the required centre of the ellipse. One of the needle points, which projects slightly from one of the major axis legs, is put over this mark, and the other needle point is put on the line beyond the position in which the ellipse is to be produced. The three inches here given is an arbitrary quantity; presuming the points in the major axis of the instrument are six inches apart, it would of course vary with the size of the instrument; the best way is to have a small ivory scale packed in the case, upon which the distance from one point to the centre of the instrument is marked.

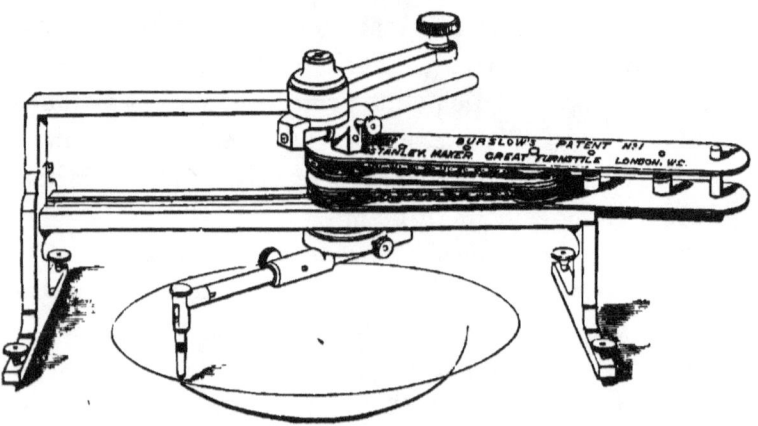

Burstow's Patent Elliptograph.

The ELLIPTOGRAPH illustrated above is the invention of Mr. Edward Burstow, of Horsham, Architect. In some respects it is the best, but in construction it is somewhat difficult, and, as at present made, expensive.

The principle of the instrument is somewhat the same as that last described—that is, a combination of a circular and rectilinear motion; but the construction is carried out in a very different manner. Instead of a circular disk and slides, this has a regular axis for the circular motion and a kind of link motion for the rectilinear. The apparatus is somewhat complicated; the following description is taken from Mr. Burstow's specification :—

"It consists of a stand supporting a horizontal frame or bed, from one end of which rises a fixed arm which terminates over the centre of the instrument in a head or socket, through which passes a vertical spindle, terminating above in a handle by which motion is given to the apparatus. The lower part of this spindle carries a horizontal bar to a slide and clamp, on which is attached a pin or axle, which travelling in a circle gives motion to a wheel and an arm. In the bed above mentioned work two slides, which may either work separately or one within the other, as hereinafter described. The principal or larger slide is about equal in length to the bed, and has near its centre a slot, or parallelogram-shaped opening, within which works a second slide, through the centre of which passes a horizontal bar, to which is attached a pen or other instrument, and the upper extremity of which terminates in a wheel. From or near one extremity of the moveable slide, rises a vertical pin or axle, on which moves freely a horizontal arm. This arm is pierced by two other spindles or pins, one near the free extremity, the other at its centre, also carrying a wheel, and forming a joint on which moves another arm, the other end of which is attached to and turns on the spindle passing through the moveable slide. To each of the three spindles mentioned—namely, that in the moveable slide, that at the centre of the arm, and that at the free end of the arm—is attached a wheel, and these wheels are connected together by chains. The wheel at the free end of the arm is turned by the spindle or axle operated upon by the bar connecting it with the milled head or handle above, and communicates motion to the wheel in the centre of the arm, which again communicates motion to the wheel on the centre of the slide, thereby giving motion to the pen or other instrument connected therewith; and thus, by the combined action of the arms, wheels, slides, and chains, causing the pen to produce an ellipse."

This instrument, unlike all other elliptographs, will produce the ellipse in either direction to the axis, and its eccentricity or distance of foci is only limited by the entire length of the instrument; thus the ordinary eight-inch instrument will produce an ellipse of from fourteen inches by seven inches to one of half an inch by a quarter.

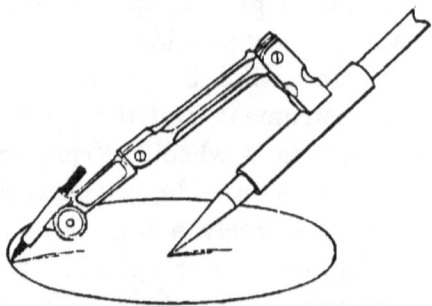

Elliptical Additions to Triangular Compasses.

A simple but imperfect form of elliptograph is made by placing a sliding fitting on one leg of the triangular compasses, shown further on. Upon the sliding fitting is jointed a short bar, which carries a pencil. The bar is also jointed in its centre, so as to adjust the point.

The ellipse is produced by placing the leg of the triangular compasses as an angle to the surface equal to the elongation of the ellipse, from the circle which it would form with the leg erect, and in this position moving the pencil so that the sliding fitting can slide on the bar as it moves. The pencil is constantly oblique to the surface, and makes its mark from different points upon the lead; therefore the ellipse is so far imperfect. The instrument is also somewhat awkward to use, but as elliptical instruments are generally expensive, it forms a cheap addition to the ordinary form of triangular compasses with moveable bar. Where the pencil line is produced, a clear line may be drawn over it in ink, by use of a French curve in parts of it that will fit.

There are many other instruments for striking ellipses, a description of which would occupy much space; most of them appear to the author imperfect in principle or of difficult construction; some do not hold the pen erect, some work on oblique sliding planes, some with cog-wheels, some guided by curves, etc.; indeed, the construction of instruments for striking ellipses seems to have been a favourite theme for exercising the fancy of inventors. Unfortunately, from the many mistakes made in the construction of this kind of instrument, many draughtsmen look with contempt upon all, considering them impracticable. The use of a correct instrument must, nevertheless, be very great, if we consider the many purposes to which it is applied,—as representing the circle in perspective bridges, skew-arches, etc.

CHAPTER XI.

INSTRUMENTS TO PRODUCE SPIRAL LINES— HELICOGRAPHS.

SPIRAL LINES are occasionally required in mathematical drawing, for delineating springs, turbines, Ionic capitals, etc. They are generally produced imperfectly, although, perhaps, sufficiently exact for practical purposes, by quadrants of circles whose centres are eccentric to the axis of the spiral; that is, the centres are placed in such a position that each consecutive quadrant is joined to the last produced, and is of a diminished radius. There have been many contrivances for producing the lines to greater perfection, a notice of which can scarcely be omitted in a work of this class; although, perhaps, there is not one of these schemes which will produce its forms sufficiently under control to be of great practical use to the draughtsman.

The object to be obtained in every instrument of the kind, or HELICOGRAPH, as it is called, is to provide apparatus or appliance by which the marking point may advance or recede to or from the centre, in regular proportion, either by equal degrees, or by accelerated motion during the revolution of the instrument. Several methods have been employed to effect this: one of these, which answers moderately well, is to draw the marking point inwards by means of a screw which forms the axis of a milled-edge wheel. The wheel revolves by contact with the surface of the drawing paper during the revolution of the instrument, the marking point being fixed upon a kind of carriage, which slides upon a bar, so that it remains steady during the motion of the screw. As spirals are required to coil both to the right and to the left hand, two screws and fittings are necessary in each instrument of corresponding right and left handed threads. The edge-roller which carries the screw gives rate of motion to the screw according to the distance it is clamped from the centre of the instrument, the rate being increased by the greater distance, causing the greater number of revolutions, and consequently the more rapid coil of spiral. This instrument will produce very small spirals in geometrical proportion : its faults are the complication of the fittings and changes of parts, and the difficulty of producing exactly required results.

In another helicograph* the point moves towards the centre with the revolution of the instrument, by means of a pair of pulleys, one of which is fixed stationary upon the axis, and the other fixed to revolve at a distance from it. A line passing over the two pulleys is attached to the marking point, which is carried along by its action. This point is fixed in a small carriage, that slides upon a light bar, by which it is steadied and directed in a line towards the centre. The pulleys are grooved cones, to give variety in the rate of eccentricity. The instrument is rather difficult to use, and the centre is encumbered.

The instrument illustrated † is, perhaps, the best kind of helicograph for producing spirals of large or of moderate size, the whole of its working parts being complete

---

* Illustrated in Adams' " Geometrical and Graphical Essays," 1791.

† Registered 1850, by Messrs. Penrose and Bennett.

in one instrument, without any detached pieces, and its movements being easily set by an exact scale which is engraved upon the instrument, so that its action is entirely under the control of the draughtsman to produce the required result. The principle of its action is that the marking point of the instrument is carried to or from the axis, during its revolution, by the motion of a thin milled-edged roller or wheel, which may be placed at any degree of obliquity. The tendency of the wheel is to travel perpendicularly to its axis; therefore, to give the required motion, the carriage in which it is fixed is loaded, to produce sufficient friction upon the surface of the paper to carry the point in the direction of its obliquity.

Penrose and Bennett's Helicograph.

The principal details of the construction of the instrument are as follows:—The *axis* is a needle point, which slightly punctures the paper, and forms the centre or axis of the intended spiral. This is placed upon a short arm, and is attached to the cylindrical·rod which forms the beam of the instrument. This rod is supported parallel with the surface of the drawing by a frame upon castors. Upon the rod a kind of carriage, sliding loosely in two bearings, is placed, in which is fixed the thin wheel that produces the oblique action. The wheel is centred in a hollow disk, and has a rack motion to move it to any angle, the inclination of

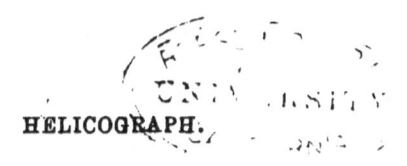
which may be ascertained by the divisions engraved upon the carriage. The marking point is fixed near the centre of the disk, in a line with the axis.

To use the instrument after it is set to the required rate or angle, it is merely necessary to move the bar round about the centre or axis, and the carriage with the marking point will follow the inclination of the rolling wheel, and produce the required spiral.

This helicograph produces its lines with considerable accuracy and beauty. The defects of it are that it will not draw a spiral except near the centre of a moderately large drawing, as the whole instrument must have room, and be supported during its revolution. From the necessary weight of the carriage to produce the required friction on the wheel, the instrument is rendered clumsy for the production of small spirals. It draws ink lines much better than pencil ones, which is a rare quality in any instrument of the class.

CHAPTER XII.

THE PARABOLA is one of the most valuable lines in mathematics and mechanics. It represents the form of surface by which the parallel rays of light, sound, heat, or other physical principle subject to the laws of reflection, may be brought to a focus, or thence diffused in parallel rays, a familiar instance of which is the reflector employed in a modern lighthouse. The parabola also represents the arch of greatest strength, approximately the path of a projectile through the atmosphere, and other important systems.

Geometrically, the parabola is shown by the outline of the section of a cone cut parallel with one of its sides. It may be produced accurately, of given proportions, in a very simple manner, by means of a *small tee-square, a straight-edge, a piece of fine cord, a pin, and a pencil,* in the following manner, of which the engraving following is an illustration. A line being drawn to represent the axis of the intended parabola, the pin, which has the cord looped over it, is pressed into the intended focus shown on the axis line; the straight-edge is then placed vertically to the axis of the intended parabola, in the position known in geometry as the directrix; it may be held firmly in this position by leaden weights. The back of the tee-square, which should be flat, and at right angles with the blade, is then placed against the straight-edge, and the edge of the blade is brought in a line with the intended base of the parabola. If the cord which is looped over the pin be now stretched to the edge of the square, and attached to it in the point where the base of the

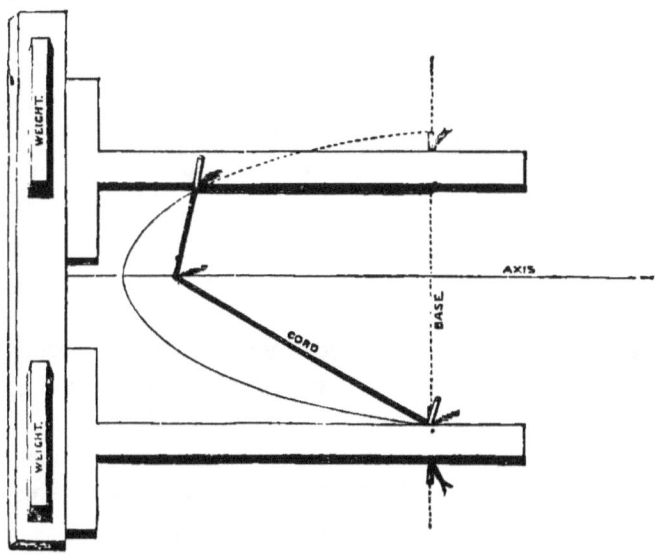

parabola cuts the edge of the square, as is shown in the lower part of the illustration, the parabola of required dimensions may be struck by drawing the cord, constantly stretched by the point of the pencil against the edge of the square, at the same time allowing the square to slide upon the straight-edge until it reaches the axial line. To prevent the cord slipping under the pencil, a notch should be cut in the side of it, close to the point, as deep as the centre of the lead. The operation of striking the parabola is shown in the upper portion of the engraving. Of course, one motion will only produce the half of the parabola, the other half will be produced by proceeding similarly with the tee-square on the reverse side of the axis. There are two squares shown in the cut, for the sake of showing two positions; one only could actually be used.

In the above description, an ordinary tee-square is mentioned; for practical purposes, the tee-square should be made specially, and be quite flat on both sides, as this will allow it to be turned over without further setting, to produce the complement of the parabola after one side has been produced. If an ordinary tee-square is used, the cord has to be tied to the edge of it; this is very awkward, especially to leave the cord of a proper length. If the square be made for the purpose, the blade should be very narrow, and have holes at frequent intervals—say, every half-inch—perforated transversely through it. The cord may be passed through one of these holes, which, by moving the straight-edge, may be made to correspond exactly with the base of the intended parabola; and when the cord is drawn through to the required length, it may be fastened by passing a small wooden peg into the back of the hole.

Separate instruments are used in the description given above, for the reason that a parabola is only occasionally required by many draughtsmen, who would not feel it worth while to purchase an expensive instrument, particularly as the above conveys the principle upon which the parabola is always struck. A complete instrument to strike the parabola could be made by merely connecting the separate pieces described by slides, and causing the pencil holder also to slide along the edge of the blade of the square, the mechanical details of which will be easily conceived without further illustration.

The HYPERBOLA is a curve shown by the outline of a

section of a cone cut parallel to its axis; it is a geometrical form much less used than the parabola.   It may
be readily struck with a *small straight rule, a pin, a
piece of cord, and a pencil,* somewhat similarly to the
parabola, except that the rule against which the cord is
drawn is fixed in or against the position of the apex

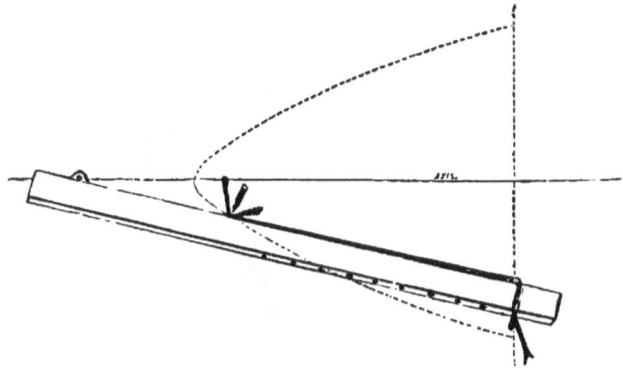

of the cone, about which it moves as a centre.   One
end of the cord is placed round a pin fixed in the focus
of the intended hyperbola, and the stretched cord is
attached to the rule at the position of the intended
base of the semi-hyperbola.   The pencil is made to
draw the cord constantly stretched against the rule
upwards, and causes the rule to follow upon its centre,
which produces the semi-hyperbola.   The opposite
reverse side is drawn by turning the rule over, thereby
bringing the centre to the opposite edge of the rule.

CHAPTER XIII.

INSTRUMENTS FOR PRODUCING CONCHOIDS, FLUTES OF COLUMNS, THE WAVE-LINE, ETC.—CONCHOIDOGRAPH.

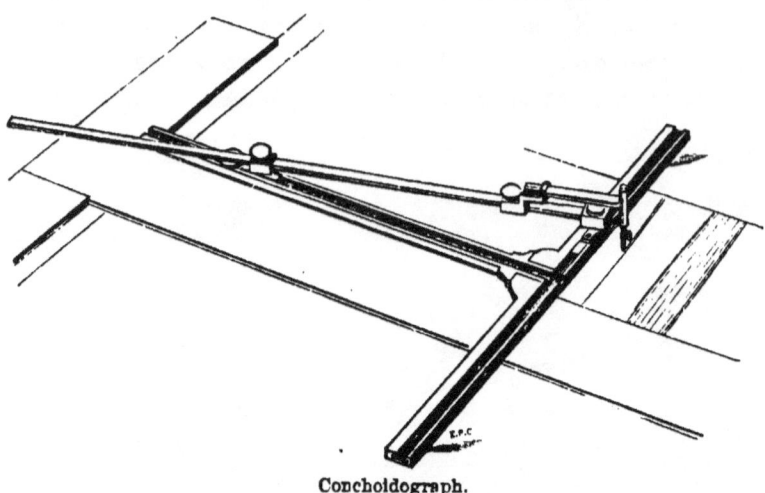

Conchoidograph.

THE CONCHOIDOGRAPH, of which an illustration is given above, is a drawing instrument constructed by the author to produce the conchoid of Nicomedes. It is intended to be used to describe columns and pilasters, with and without flutes, upon architectural drawings. It may also be used to strike the wave-line, the semi-ellipse, and for some other purposes.

The principle of the instrument is illustrated in Nicholson's "Five Orders of Architecture," in which, in speaking of diminishing columns, he says : "The best and most simple manner is by means of an instrument invented by Nicomedes to describe the first conchoid, for this being applied to the bottom of the shaft performs at one sweep both the swelling and the diminution, giving such a graceful form to the column that it is universally allowed to be the most perfect practice hitherto discovered. The columns in the Pantheon at Rome, accounted the most beautiful among the antique, are traced in this manner." The instrument that is then described by Nicholson is adapted to give the workman the true form of the full-size column upon the stone, for which a small portion only of the conchoid is required. This instrument could not be reduced to a drawing instrument without a different construction, from its very limited action, particularly from its being adapted to work to one hand only, and from the want of any system of adjustment. Nevertheless, the principle of action is so correct that it only required a little constructive detail to render it an efficient drawing instrument.

The writer's conchoidograph, illustrated at the commencement of this chapter, in detail consists of one straight metal rule, to the centre of which another similar rule is joined at right angles. Each of the rules has an under-cut groove along it, and a metal slide that moves easily in the groove. The slide in the rule which is shown in horizontal position upon the engraving, has a screw, which will clamp it to the groove where required. Above the slides, and centred upon them, are two heads, similar to beam-compass heads; these are made so as to allow a light bar to slide through them, which may be clamped to either head by the screw above. At the end of the bar is a socket to carry ink or pencil point. The instrument is supported a short

distance from the drawing upon four legs; two of these
are placed in a line through the centre of the instrument, so that by causing them to touch the edge of a
tee-square, the instrument is set at once in a vertical
position.

To use the conchoidograph for delineating columns,
it is necessary, first, to set out the vertical centre line
of the column, and from the centre line to set off the
semi-diameters of top and bottom. The tee-square is
then brought up to a line with the bottom of the
column, and the instrument placed upon it, so that the
two centre legs are brought against the upper edge of
the tee-square, and the now vertical edge of the instrument is against the centre line of the column. The
three screws attached to the sliding heads are released,
so that all parts of the instrument are quite free. The
slides and bar are brought into a line with the bottom
of the column, and the pen or pencil point placed over
the mark which indicates the semi-diameter; in this
position the sliding head nearest the pen is clamped
to the bar. The pen or pencil is then brought up to
the top semi-diameter, and the other slide moved
backwards until the pen or pencil will fall upon the
required setting. In this position the back slide is
to be clamped to the *sliding groove,* so that the head
remains *stationary,* and the bar slides through it. It
is now set, and the line may be delineated.

To draw the corresponding side of the column, the
instrument is turned over on the tee-square to the
reverse hand, without further setting. By carefully
going over it once, the instrument may be set in two
minutes for any column. The necessity for drawing
the centre and top and bottom semi-diameters, of course

relates to the first setting of the instrument only, and may be done on a separate piece of paper if required. As the instrument once set answers for all the columns and pilasters of the same dimensions, therefore the separate setting out of either the base or the cap for one only will be sufficient.

The instrument may also be made to draw all the flutes of the column afterwards, perfectly and with little difficulty. It is only necessary to keep the edge of the instrument in a line with the centre of the shaft, release the slide next the point, place the pen or pencil over the various settings for the flutes (which need only be at the bottom of the column), and afterwards to clamp the same head to the setting; the instrument will thus produce all the consecutive flutes except the centre one. When it is set to any particular flute, it will be found best to produce all the corresponding flutes on different parts of the drawing, to either hand, from the same setting of the instrument. For neatness in commencing to draw the flutes, a piece of waste paper should be placed above the column, to see that the pen draws properly before the line reaches the drawing.

To draw semi-ellipses with the conchoidograph, the same description will answer as is given with the semi-elliptograph, in a previous chapter, only it must be particularly observed that both heads are clamped to the bar to produce the semi-ellipse, whereas to produce the conchoid, the bar slides freely through the back head. The difference in lines produced by this alteration is shown in the following illustration, where the perfect conchoid is shown outside the semi-ellipse. Both lines are produced from a similar setting of the instrument, except that the bar slides through the back head

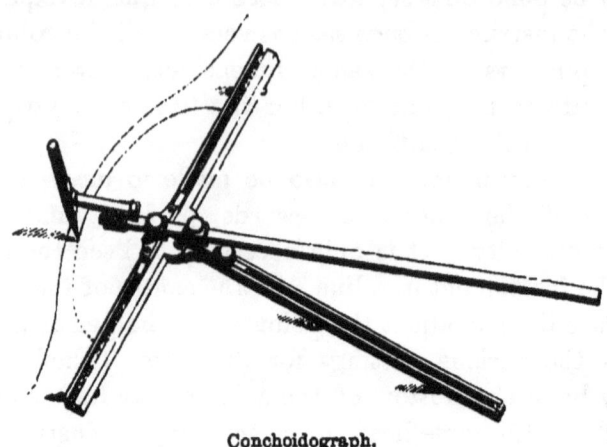

Conchoidograph.

to produce the outer figure—the conchoid; and the back head itself slides to produce the inner one—the ellipse.

CHAPTER XIV.

INSTRUMENTS FOR THE PRODUCTION OF REGULAR GEO-
METRICAL OR ORNAMENTAL FIGURES—GEOMETRICAL
PENS.

AFTER considering the many elaborate means already
described, which are at the same time the most simple
we possess, for producing ordinary geometrical forms,
we can but be struck with the simplicity of arrangement of the Geometrical Pen, which produces a thousand varieties of ornamental figures in geometrical
proportions, for the most part bearing so slight a resem-
blance to each other, that it is difficult to conceive that
they can be the production of a single instrument.

The original geometrical pen was invented by John
Baptist Suardi, who published an elaborate description
of it.* This instrument is capable of producing a
great variety of pleasing and beautiful forms, which
might be applied to design, but could scarcely be used
in drawing, as they are produced so little under the
control of the draughtsman, from want of any means
of adjustment to allow the form to be brought to any
proportional scale. Thus, each figure produced must be
of a certain size; if a larger or smaller be required, the
instrument is incapable of producing it; neither could
the figure be placed in any exact position.

Suardi's geometrical pen, it may be presumed, was
never used for practical purposes, as it is very rarely

* "Nuovo Istromenti per la Discrizzione di diverse Curve
Antichi e Moderne." See also Adams' "Geometrical Essays."

to be met with, being possibly considered more as an amusing philosophical toy than as a mathematical instrument. A few improvements in details were made by our mathematical-instrument makers, but nothing in the construction to render it useful.

It appeared to the Author, upon examining this very ingenious and curious instrument, that the geometrical principle on which it was founded, namely, the continuous evolution of every variety of epicycloid and hypocycloid, must produce useful ornamental forms if the practically necessary means of adjustment could be applied. After a few experiments, the writer was able to improve the general construction, and to produce the instrument shown in the following illustration, which materially differs from the original, particularly in the addition of the lower horizontal bar and its connections, by means of which the action of the instrument is communicated to any position in relation to the centre, and the forms produced adjusted to required size; at the same time, their variety is multiplied a hundredfold. The other improvements in the construction of the new instrument were the power of raising the marking point and the means of placing the figure produced in exact position.

The construction of the Author's geometrical pen is rather complicated to describe. The details of the various parts, it is hoped, will be understood by careful perusal of the following particulars, with the assistance of the illustration.

In the first place, the instrument is supported and held at a uniform level above the surface of the drawing by a light solid frame, which stands upon cross feet, the under sides of which are cut similar to a rasp, to prevent the instrument slipping. In the inside centre of each of the feet, exactly under the centre of the frame, is a

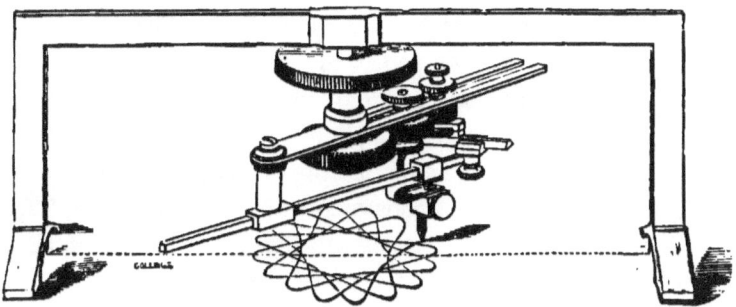

Improved Geometrical Pen.

stout needle point, which is intended just to enter the surface of the paper upon which the instrument stands. If each of these needle points be placed upon a straight line, the line will pass exactly under the centre of the axis of the instrument, which centre will also be at equal distance from each of the points. Thus, by the assistance of these points, the figures to be produced by the instrument may be placed in exact position on any line upon the drawing.

From the under side of the centre of the above described frame is fixed a stout steel pin, which forms an axis upon which the actual instrument revolves. The instrument is moved round upon the axis by a large milled head, immediately under the top of the frame. This milled head is connected by a hollow socket, which fits exactly upon the pin, but which may be raised by sliding it up the pin a sufficient distance to lift the whole instrument, and consequently with it the marking point, off the paper. Connected with the socket is a horizontal steel flat bar, which is slotted down the centre to form a bed. Upon the bed are fitted two heads, which form the centres upon which a series of three cog-wheels are placed; the heads slide upon the bed, so as to bring any two or three of these cog-wheels, as may be required, into gear.

The cog-wheels are made of various sizes, the number of teeth in each being, if required to produce geometrical figures, some multiple of a common number. The number of teeth we have selected are, 120, 96, 72, 60, 48, 36, and 38, the first six being multiples of 12.

There are three axes to the cog-wheels—the centre a fixed one, and the two moveable ones on the heads. The wheels are made changeable from any one of these axes to the other. Each of the wheels has a small key-way cut through on one side of the hole in the centre of the wheel; this enables the wheel to revolve on a plain pin, or to be fixed when placed upon a pin with a projecting key.

The action of the wheels is as follows. At the centre or axis of the instrument is a keyed pin : any wheel placed upon this is fixed, and does not revolve with the other portion of the instrument; thus this wheel gives action to the others that revolve round it. The axis of the second wheel from the centre is a plain pin ; therefore a wheel upon this revolves with the action of the first, and communicates its action to the third. The pin of the third wheel from the centre is keyed; therefore this wheel and axis both revolve. Upon the axis of the third wheel is fixed a moveable horizontal crank piece, which forms a bed upon which a sliding head may be clamped in any position. This sliding head forms a centre, and gives motion to the horizontal bar along the under part of the instrument, the other end of which bar is carried by a sliding fitting that allows it free horizontal action in any direction. A small head upon the bar carries the ink or pencil point, which may be clamped in any required position.

The above-described bar, as has been already mentioned, does not form a part of former geometrical pens. The original principle was to place the marking point upon the crank piece; consequently, the figures it produced were uniformly somewhat larger than the circumference described by the third wheel around the principal axis of the instrument.

To produce variety of geometrical figures by the arrangement and setting of the various parts of the instrument, the following rules may be generally observed : That the changing of the second wheel for any other will make no material difference in the figure. That by placing one of the smaller wheels upon the centre axis, and one of the larger wheels upon the third axis under the crank, figures composed of spiral forms inscribed in a circle, similar in character to Nos. 1 and 2, will be produced. That by placing one of the larger wheels upon the centre axis, and one of the smaller upon the third axis, external foliated or leaf-like forms, similar to Nos. 3, 5, 6, 7, 8, and 9, will be produced.

By the eccentricity of the crank, the amount of deviation from a circle is produced; thus, if the head which gives action to the bar be clamped over the lead centre of the third wheel, a plain circle will be produced; if the head be clamped slightly eccentric, an annular interwoven figure, similar to No. 4, will be produced; if the action from the crank be very eccentric, foliated angular figures will be produced similar to Nos. 3, 5, and 9.

Figures drawn by Geometrical Pen.

By the movement of the ink or pencil head upon the horizontal bar, the figures will be entirely varied. Thus, if the head be near the action of the crank, the figures will be gradual curves, as No. 4. If the head passes with the action of the bar to the centre, the figures will be angular, as Nos. 3, 5, and 9. If the head be clamped beyond the centre, as far as possible from the crank, the figures will be looped, and the number of foliations around the centre doubled, as Nos. 7 and 8.

The above are the general laws; of course intermediate settings produce intermediate forms, until one passes into the other.

The number of points or leaves of an external foliated figure will be produced by the arrangement of the wheels. The rule is, to consider the number of teeth in one wheel to the number in the other as a vulgar fraction, of which the centre wheel is the denominator, and the third wheel, the one that gives action to the crank, the numerator. If the fraction be reduced to the lowest terms, the *denominator* will give the number of points or foliages. Thus, if we put the 120-wheel on the centre, and either a 96, 72, or 48-wheel on the third axis, a five-leaved figure will be produced, because—

$$\frac{96}{120} = \frac{4}{5} \qquad \frac{72}{120} = \frac{3}{5} \qquad \frac{48}{120} = \frac{2}{5},$$

the denominator in each instance being a five.

The following numbers will give one example of each regular foliated figure, produced by the wheels we have selected; the number of foliages will scarcely be distinguished in some of the illustrations, by reason of the variations given by other means already described.

120 centre	60 crank	2	leaved figures, ovals, example	No.	2.
72 ,,	48 ,,	3	,,	,,	,,	1.
120 ,,	30 ,,	4	,,	.,	,,	7.
120 ,,	72 ,,	5	,,	,,	,,	8.
72 ,,	60 ,,	6	,,	,,	,,	6.
96 ,,	36 ,,	8	,,	,,	.,	5.
120 ,,	36 ,,	10	,,	,,	,,	3.
72 ,,	30 ,,	12	,,	,,	,,	4.

It may be observed, that the nearer these figures are to like numbers, the more rounded the foliages will be, other particulars considered. This may be observed

in No. 6, where the wheels are 72 and 60; and the
reverse in No. 3, where the wheels are 120 and 36.

If the figure produced by any setting of the instrument is required to be enlarged, the head upon the
crank which gives action to the bar should be moved
away from its centre one or more divisions, and the ink
or pencil head moved one or more divisions nearer the
crank. If the figure is to be reduced, this should be
reversed. The divisions upon the crank piece and the
horizontal bar cannot be exact proportions with all the
wheels, but they assist in moving the heads approxi-
mate distances.

To keep the instrument in good order : after using it,
the steel portions should be wiped with a piece of wash-
leather, previously moistened with oil ; this may be kept
in the case with the instrument.

CHAPTER XV.

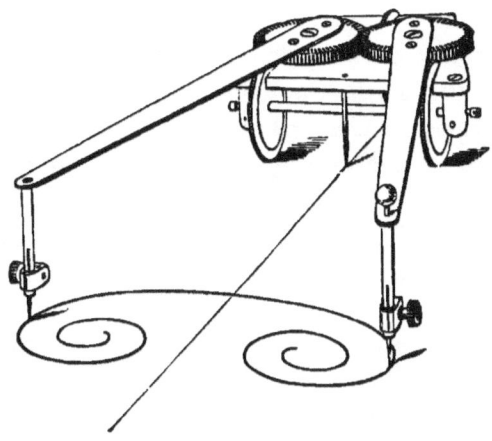

THE ANTIGRAPH is the only instrument that has ever
been invented or made public—known to the writer—
for drawing parts of a figure the reverse hand to the
original. It performs its work very nicely, but practically it has very limited use. It is introduced here
to complete the series. It consists of a carriage supported upon three moderately large wheels, two on
one axis, and the other upon its own axis. The axes
run lightly between centres; and the wheels have
very thin edges, so as to give a slight bite upon the
paper, to enable them to run perfectly straight and
parallel. The single wheel is placed in the centre or

H

axis, and a corresponding point descends from the front of the carriage nearly to the surface of the paper. If the centre wheel and point are placed on a straight line, the carriage will move continually over this line. Upon the top of the carriage are placed two equal cog-wheels, made to run softly together without shake. Upon the surface of the wheels are jointed, or lightly sprung, two equal arms, which extend out some distance in the front of the carriage, and carry the pencil and tracing point, which are exchangeable from one arm to the other. It is manifest that any motion given to one of the arms will be exactly communicated to the other in the reverse direction, and as the carriage keeps exactly in one line throughout all movements, equal motion will be given to each arm.

To use the Antigraph. Draw a line, which is to be the geometrical axis of the drawing. Any figure or part sketched upon one side of the line, as one side of a vase or capital, may be traced off on the other. If a sketch is required to be copied to the reverse hand, it may be pinned down upon the face of the drawing paper and traced upon it.

The more general means of copying reverse parts of a geometrical drawing is by taking a tracing of one side of the drawing upon tracing paper with a fine H.B. black-lead pencil, the lines being made sufficiently clear to be seen distinctly upon the back side. Now by placing a sheet of black-lead paper over the intended reverse part or opposite-handed drawing, and pinning the tracing the back side upwards over this, the whole of the lines may be gone over with an agate tracer, and a fine black-lead drawing produced, which only needs touching up and correcting by the eye.

This is also a most delicate method of transfer, and is much used by some of our best artists in illuminating.

For producing equal reverse parts for rougher details, as for joiners and masons' work, the single sheet, with design, may be folded in the axis line with the drawing outwards, and the transfer of equal reverse parts made on the back by placing a sheet of doubled carbonic paper in the fold, and going over the sketch or drawing with a tracer. This will produce the lines to the right and left hands in an indelible ink, which will be afterwards clear to work to.

Where single reverse curved lines are required upon drawings, equal ordinates from each side of a perpendicular will give points through which the required curve may be traced with sufficient accuracy for practical purposes. If there are many such lines, dividing the entire surface into equal squares is the most expeditious means, and tracing through points set off on these squares as required for the line positions.

CHAPTER XVI.

INSTRUMENTS FOR COPYING DRAWINGS—TRIANGULAR COMPASSES—TRIANGULAR BEAM COMPASSES, ETC.

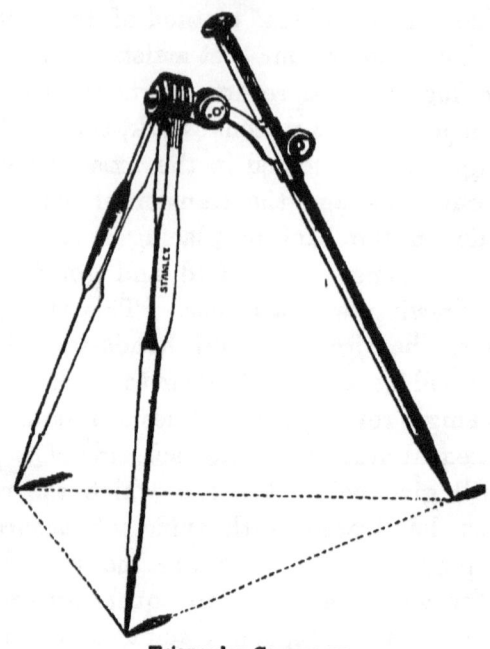

Triangular Compasses.

TRIANGULAR COMPASSES are principally used for copying plans of land or for fine-art drawings, for which purposes they are very convenient, as they will give the exact position of one point in relation to two others; they are also useful to test the accuracy of the copies of plans.

The ordinary form of triangular compasses is in construction similar to a pair of dividers, having in addition a third leg jointed upon the centre pin of the head joint. This leg moves round horizontally with the pin, and vertically upon its own joint, thus producing a universal motion.

Another plan of making the universal jointed leg, which is that illustrated, is to carry over a joint and socket from the centre pin, instead of jointing the leg directly upon it. The third leg is made a plain pointed rod longer than the compasses. The socket has a clamping screw attached to it, which enables this leg to be clamped out longer than the others if required ; it also admits of the convenience of a plain black-lead pencil being fixed in the socket in place of the rod, which, if rather less exact than the plain point, is much more convenient when employed for copying ordinary drawings.

In using the triangular compasses to obtain the position of one point in relation to two others, it is customary, first, to set the two legs which form the legitimate dividers over two given points, and afterwards to move the universal leg to the third point.

The triangular compasses may be used as dividers without employing the third point, which saves the necessity of shifting the instrument in copying.

The Author invented a very simple kind of triangular compasses for copying large plans. This instrument is made similar to the plain beam compasses, except that in the middle of the beam there is constructed a kind of knuckle rule-joint, the centre of which is carried down to the length of the points on the beam heads, and is made to form the third point. In this instrument the points are always erect, which gives more exact position than when they are held obliquely, as in

the first-described kinds.   These compasses may also,
by unscrewing the point in the joint, be used as ordinary beam compasses; or the common beam compasses

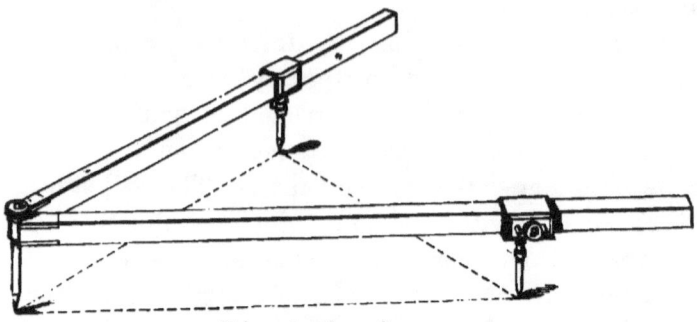

Triangular Beam Compasses.

may be altered into this kind of triangular compasses
by having a joint and point fitted to the centre of the
beam, which may be done without otherwise affecting
its ordinary use, at the same time rendering it more
portable.   Two-feet beam compasses of this kind may
be packed in the ordinary magazine case of instruments,
and will be found useful, if not used for triangulating,
for producing circles and taking off distances within
its dimensions.

CHAPTER XVII.

INSTRUMENTS PRINCIPALLY USED FOR ENLARGING OR
REDUCING DRAWINGS—WHOLES AND HALVES—PRO-
PORTIONAL COMPASSES—PROPORTIONAL CALLIPERS.

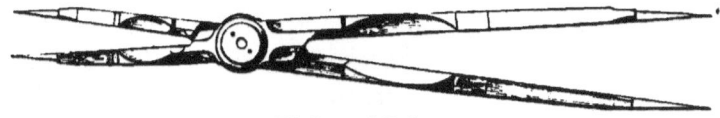

Wholes and Halves.

WHOLES AND HALVES, as illustrated above, are so
called because, when the longer legs are opened to any
given dimension, the shorter ones will open to half of
that distance. They are used principally for bisection,
as a whole quantity being taken from any scale with
the long legs, it may be set off equally on each side
of a given line with the shorter ones, or the centre
between two points may be instantly found. They are
also occasionally used to make details double dimensions, or plans half dimensions.

In construction, they consist of two pairs of compasses, with one common joint, which is placed at one-
third of the entire length distant from one of the pairs
of points. Each of the legs of one of the long pair of
compasses is connected with one of the legs of the
shorter pair, through the joint, which it crosses in
passing. They are a difficult instrument to make, from
the reason that one of the points has to be soldered to
the joint after it is otherwise finished.

Wholes and halves are not much used, but they

have some advantages, for their peculiar purposes, over proportional compasses, which have superseded them, as they are a much more convenient instrument to handle, and the points may be set fine.

A kind of proportional compasses has been introduced, with knuckle-joints in the centre of the long legs of a pair of compasses otherwise similar to the wholes and halves; the joints allow the points to be turned down sideways, at right angles with the head joint, thus allowing the longer points to be brought nearer to the centre of action, and thereby alter the proportions to a limited extent.   They are very clumsy to use, as expensive as proportional compasses, and altogether inferior.

Proportional Compasses.

PROPORTIONAL COMPASSES are principally employed for reducing or enlarging drawings in any given proportion, either superficial or solid.   They may also be used for the division of the circle into equal parts, and the extraction of the cube and square root of a dimension.

The illustration above represents the most simple form of this instrument, which is also that most generally used.   In detail, it consists of two narrow, flat pieces of metal, each having a dovetail groove up the centre for the greater part of its length, and a steel point at each end.   These two pieces, called sides, are united by a pair of slides, fitted together upon one pin and also fitted in the grooves, so that they will slide along them.   A milled-head screw clamps them together upon the pin, which forms the axis of the compasses. There is a stud upon one of the sides, and a corresponding notch in the other, which brings the points over each other when the instrument is closed, and prevents any shifting of the sides, in moving the slides to set the instrument.

Proportional compasses most frequently have scales divided on the four plain surfaces, by the sides of the grooves, which are respectively engraved with the words—Lines, Circles, Plans, and Solids.

The scales are read off when the instrument is closed, by bringing the line upon the slide opposite the required division. They are used as follows :—

The Scale of Lines is used to reduce or enlarge drawings in giving proportions. The line on the slide being set opposite to the line under any figure on the scale, the proportion of the opening of the instrument will be as that figure is to 1 ; for instance, if a drawing is required to be made either one-third or three times the dimensions of the copy, the line on the slide is put to the 3 on the scale of lines, and the slide is clamped; if the compasses be then opened, one pair of points will prick off three times the distance of the other pair.

The scale of lines being only in proportion of some figure to 1—as 4 to 1, 7 to 1, etc.,—it will strike the draughtsman as being of limited use, as the frequent reductions required bear no proportion to 1. For instance, a plan at 3 chains has to be reduced to 5 chains or enlarged to 2 chains, the proportion being as 3 to 5 and 3 to 2 respectively. In these instances the slide has to be shifted about until the points correspond with the two scales required. The line of lines is unnecessarily defective in this respect in the part of the proportional compasses mostly used, and there are spaces quite unoccupied where useful proportions might be put in, as 2 to 3, 3 to 4, etc.

The Scale of Circles is used to divide the circumference of a circle into any number of equal parts, up to 20. The slide being put to the number of divisions required when the instrument is closed, to whatever radius the long points are opened, the shorter points will divide the circumference of the circle struck by it into that number of equal parts, according with the setting. This scale of the proportional compasses is never used in practice, and is sometimes omitted: the points of the compasses are too obtuse to divide a circle nicely.

The Scale of Plans is employed to reduce or enlarge the area of a plan in a given proportion; for instance, if the line on the slide be set to the 5 on the scale, a circle struck with the long points will inscribe a surface five times the area of one struck with the shorter points; of course, any other figure will follow the same rule.

The Scale of Solids is used to reduce or enlarge the contents of a solid in a given proportion; for instance, a set of drawings of a gasometer being given, and another set of drawings required, in the same proportion and to the same scale, of five times the capacity, it would be necessary to set the slide to 5 on the scale of solids; then every dimension taken from the original drawing with the short points of the proportional compasses would be set off on the drawing to be produced by the longer points.

The draughtsman may work out many other uses from the proportional compasses, particularly from the scales of plans and solids, as the numbers on their respective scales are the squares and cubes of the ratios of the lengths of the opposite ends of the compasses. Thus, in practical mechanics, from drawings to scale of known examples of resistance, strain, velocity, percussion, etc., other drawings may be produced to the same scale, with different dimensions, proportions, etc., without the necessity of calculation.

For general purposes, 9-inch proportional compasses will be found more useful than 6-inch ones.

Bar Proportional Compasses.

Proportional compasses are very frequently made with an adjusting screw, as represented in the illustration. This enables the slides to be set exactly to the division; it will also, by fixing the adjusting screw on a small stud provided for the purpose, adjust the points in the manner of hair dividers. This arrangement is sometimes useful, but it renders the instrument more clumsy and expensive. It can scarcely be recommended.

The French method of adjusting the slide, by a rack within one of the grooves and a pinion upon the slide, is less clumsy than the English method, although perhaps not so durable. Altogether, by comparison, its merits appear greater than its defects.

In the selection of a pair of proportional compasses,

it may be observed that it is very important that the slides and centre-pin fit at all parts of the grooves, as a small shake on the centre will considerably alter the proportions.

There is a particular drawback to the more extensive use of the proportional compasses—that is, that they can never be *set* to keep the points sharp, as may be done with the ordinary compasses; as setting would entirely destroy the proportions of all the divisions; thus it is necessary to make a very strong obtuse point, which is very difficult to be seen when it is placed on the drawing.   This very great defect may be

remedied by constructing the instrument with all the points turned down, about half an inch, at exactly right angles with the sides of the instrument, similar to the point illustrated.   If the points are made in this manner, keeping the outer edge of the point true with the division, they may be made sharp and fine, and can be set at any time from the inside, without altering the divided proportions.   Another defect is, that in ordinary proportional compasses the points slide sideways over each other; this can be remedied by placing the points edgeways upon the sides, which causes them to meet as in ordinary compasses.

The writer has lately made a model pair of proportional compasses for Mr. Oliver Byrne, the editor of Spon's Mechanical Dictionary, the invention of that gentleman; although a little costly in construction, it is no doubt the most perfect instrument of the class. It is similar to the ordinary proportional compasses in principle, only that the slide which carries the centre, instead of being made to fix in proportions of from one to ten only, passes the whole length of the instrument; therefore it can be brought quite over one pair of points, so that in opening the opposite pair the first will not move. The scale placed upon the instrument divides the whole decimally, so that the slide fixed to any decimal fraction will open in this proportion to the whole, and will thus work out proportions in the same manner as the eidograph, described in the next chapter. Mr. Byrne has tables written for the instrument, which he proposes publishing. These give the proportional compasses entirely new uses, applicable to settings for a variety of purposes,—as division of wheels into any number of teeth by opening to the radius, proportions of metres to feet for changing drawings from English to French scales or the reverse, or other scales, and the working out of various useful tables.

Proportional Callipers are in every way similar to proportional compasses, except that in place of the long sharp points—a pair of mechanic's calliper points are constructed; they are used to transfer diameters of turnings, etc., from a drawing to the actual work, and may be considered more as a tool than a drawing instrument.

CHAPTER XVIII.

INSTRUMENTS FOR REDUCING AND ENLARGING DRAWINGS OF CONSIDERABLE SIZE—THE PANTAGRAPH—THE EIDOGRAPH.

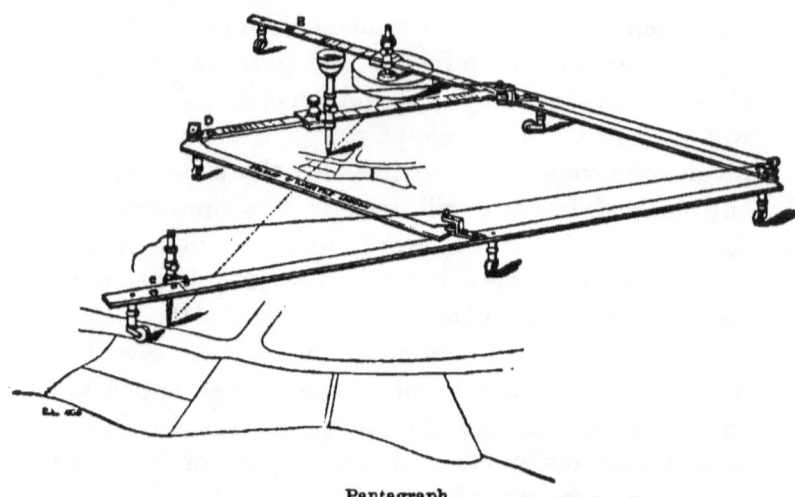

Pantagraph.

THE PENTAGRAPH, or more properly Pantagraph, is used almost entirely for copying or extracting portions of plans of land to a reduced scale. Occasionally it is used to trace in ornaments from details, and for other kinds of reduced drawing. It may also be used to enlarge drawings, but this it does so imperfectly that it cannot be recommended for the purpose.

The pantagraph, as represented in the illustration above, consists of four rules of stout brass, which are jointed together in pairs, one pair of rules being about double the length of the other. The free ends of the shorter pair are again jointed to the longer in about the centre. It is important that the distance of the joints on each of the short rules should exactly correspond with the distance of joints on the opposite longer rules, so that the inscribed space should be a true parallelogram. To enable the instrument to work freely and correctly, all the joints should be perfectly vertical, and with double axes. Under the joints casters are placed to support the instrument, and to allow it to move lightly over the paper. One of the long rules engraved (c) on the instrument has a socket fixed near the end, which carries a tracing point when the instrument is used for reducing. The other long rule (B), and one of the shorter rules (D), have each a sliding head fitted upon it, which is similar to one of the heads of a pair of beam compasses. Each head has a screw to clamp it in any part of the rule, and carries a perpendicular socket, which is placed over the edge of the rule in a true line with the joints. Each socket is adapted to hold either a pencil holder, tracing point, or fulcrum pin, as may be required. The rules upon which the heads slide are divided with a scale of proportions : 1—2, 11—12, 9—10, etc., which indicate as one is to two, as eleven are to twelve, as nine are to ten, etc.

A loaded brass weight, which firmly supports a pin that fits exactly into either of the sockets, forms the fulcrum upon which the whole instrument moves when in use.

The pencil holder is constructed with a small cup at the top, which may be loaded with coin or shot to cause the pencil to mark with the required distinctness.

Arrangement is made to raise the pencil holder off

the drawing. This is effected by a groove down one side of the pencil holder, in which a cord is fixed, passing from the pencil along the rules, turning the angles over small pulleys, and reaching the tracing point, where it may be readily pulled by the hand to raise the pencil. This will be found especially convenient when the pencil is required to pass over any part of the copy not intended to be reproduced.

The pantagraph is set to reduce drawings in two ways, termed technically the *Erect manner*, and the *Reverse manner*. It will be necessary to give full details of each manner, particularly in relation to the scales engraved on the instrument, which are not very intelligible; indeed, comparatively few professional men are sufficiently acquainted with them to avail themselves of their full value, neither is the information required given in any one of the many published descriptions.

By the *erect* manner of setting the pantagraph, the reduced copy will appear erect; that is, the same way as in the original. The general position of the parts of the instrument set in this manner is shown in the cut at the commencement of the chapter, where it will be seen that the fulcrum pin is placed in the socket of the sliding head upon the outside long rule engraved (B) and the pencil holder in the socket upon the short central rule (D). By this method of setting the instrument, it will reduce in any of the given proportions not exceeding half size, technically from 1—2. The scales engraved upon the rules that accord with the erect manner of setting are those which have 1 for the first proportion; as 1—2, 1—3, 1—4, etc. The other scales may be used, but will not accord with the reading, except through arithmetical deductions, the results of which may be given more clearly by the following complete table than by rules with exceptions.

Table of Reductions by the Pantagraph in the erect *manner, the fulcrum being placed in the outside socket upon the rule* (B), *and the pencil central upon rule* (D).

Reading given upon the scales.	Reduces in the proportion of	Reading given upon the scales.	Reduces in the proportion of
1— 2 ...	... 1 to 2	2— 3 ...	... 2 to 5
1— 3 ...	... 1 ,, 3	3— 4 ...	... 3 ,, 7
1— 4 ...	... 1 ,, 4	4— 5 ...	... 4 ,, 9
1— 5 ...	... 1 ,, 5	5— 6 ...	... 5 ,, 11
1— 6 ...	... 1 ,, 6	6— 7 ...	... 6 ,, 13
1— 7 ...	... 1 ,, 7	7— 8 ...	... 7 ,, 15
1— 8 ...	... 1 ,, 8	8— 9 ...	... 8 ,, 17
1— 9 ...	... 1 ,, 9	9—10 ...	... 9 ,, 19
1—10 ...	... 1 ,, 10	10—11 ...	... 10 ,, 21
1—11 ...	... 1 ,, 11	11—12 ...	... 11 ,, 23

In the above table the readings which agree with the proportions are given to show clearly which proportions agree with the erect scales; many of those that do not agree with the reading are very useful, as 2—3, which is often required to reduce a drawing from a scale of 20 to one of 50.

In the *reverse* manner of setting the pantagraph, the reduced copy appears reversed, or upside-down, to the original. The fulcrum pin is placed in the socket upon the short central rule (D), and the pencil holder is placed in the outside socket upon the rule (B). This is generally the most convenient way of using the pantagraph for large drawings, as the original and copy come edge to edge, and need not overlap each other, which is often compulsory in the erect manner; the range of scale is also much greater, as the proportions include the unit proportions of the erect scale and continue in ratios up to full size.

I

The following table will give the readings of the instrument which accord with the reverse setting, and those which may be used to this setting, obtained by calculation.

Table of Reductions by the Pantagraph in the reverse manner, the fulcrum being placed in the central socket on the rule (D), and the pencil in the outside socket upon rule (B).

Reading given upon the scales.	Reduces in the proportion of	Reading given upon the scales.	Reduces in the proportion of
1— 2 ...	1 to 1, full size.	2— 3 ...	... 2 to 3
1— 3 ...	... 1 to 2	3— 4 ...	... 3 ,, 4
1— 4 ...	... 1 ,, 3	4— 5 ...	... 4 ,, 5
1— 5 ...	... 1 ,, 4	5— 6 ...	... 5 ,, 6
1— 6 ...	... 1 ,, 5	6— 7 ...	... 6 ,, 7
1— 7 ...	... 1 ,, 6	7— 8 ...	... 7 ,, 8
1— 8 ...	... 1 ,, 7	8— 9 ...	... 8 ,, 9
1— 9 ...	... 1 ,, 8	9—10 ...	... 9 ,, 10
1—10 ...	... 1 ,, 9	10—11 ...	... 10 ,, 11
1—11 ...	... 1 ,, 10	11—12 ...	... 11 ,, 12

The above table and the previous one give the proportions for reductions, the tracing point being in every instance considered as placed in the socket upon the rule (c). If it were required to produce an enlarged copy, which the pantagraph will do but very imperfectly, the pencil and tracer would have to change places; the proportions of course would read the same.

In using the pantagraph some care is required in setting the fulcrum weight in the best position to allow easy action of the instrument over the space required. It should always be roughly tried over the boundary before commencing the copy.

The ordinary pantagraph will in no instance work over a large drawing at one operation, but it may be shifted about as required, using care, and testing the copy after the fulcrum is moved, to see that the tracer and pencil correspond in those parts already produced, that the pantagraph will reach in its shifted position. The fulcrum weight, being generally made with needle points to attach it to the drawing, will be found very difficult to shift so short a distance as is frequently required. This may be easily remedied by attaching with gum a piece of india-rubber over each of the sharp points, when it is required to be used for large drawings. The rubber will hold the paper sufficiently if the pantagraph work freely in the joints and castors, as it should do.

In copying buildings which frequently occur in plans of estates, etc., a straight slip of transparent horn will be found very convenient to guide the tracing point. Some draughtsmen have the horn cut with an internal angle, by which one side and one end of a building may be traced without shifting the horn.

Architects and mechanical engineers seldom use the pantagraph; however, it may, perhaps, be sometimes used with advantage for tracing in the most difficult and tedious parts of a drawing with a precision impossible by the hand. This applies particularly to such parts as are frequently repeated, as capitals, trusses, bosses, tracery, etc., upon drawings to very small scales. In these instances it is only necessary to make a detail sketch, say six times the size required, and to place the fulcrum weight in such position as the pencil will pass over the parts required to be filled in, the tracer at the same time resting on a corresponding part of the detail sketch, which may be placed in position under the tracing point, and be held sufficiently by two lead weights. For a second ornament on the same drawing, the detail may be shifted without moving the fulcrum.

To follow the outline of any object of the ornamental

class, or for the reduction of mechanical drawings to
a size suitable for wood or other engravings, the strip of
horn will be found particularly useful; indeed, to obtain
any degree of precision in geometrical figures, it will
be better, generally, to let the tracer follow a guiding
edge placed over the original for the purpose.   French
curves are particularly useful, although perhaps only a
small piece may be available at once.   The tracer may
rest on the surface until another part of the curve is
found to correspond with the continuation of the line.
The author has made use of this plan for drawing some
of the illustrations in this work upon the wood.

In some old pantagraphs a guide is fixed to the
tracing point.   The guide is a kind of handle similar
to a drawing pencil, the point of which is hinged to the
point of the tracer.   This gives a convenient and firm
hold of the point, and appears to the author a useful
appendage.

Pantagraphs have been made in many shapes unnecessary to describe, as they are all of one principle
—that of a parallelogram jointed at the four corners;
the principal difference being in the position of the
points and fulcrum in relation to the parallelogram.
One thing is essential in every construction,—that is,
that the fulcrum, tracer, and pencil should always be in
a true line when the instrument is set for use.   The
parallelogram may be in any position on the instrument, to the fancy of the maker.

For certain uses of the pantagraph a diagram of the
principles of its action may be useful.

Thus, if A B, B C be equal lines, it will be quite clear
that if we let a perpendicular fall from the angle B,
upon the dotted line A C, it would bisect that line.

Now, if we bisect the lines A B and B C in E and F, and erect a triangle equal and similar to the portion of the first E B F, if this is placed opposite to the first, as at E D F, the point D of this would also bisect the dotted line A C. Now supposing all the angles of this

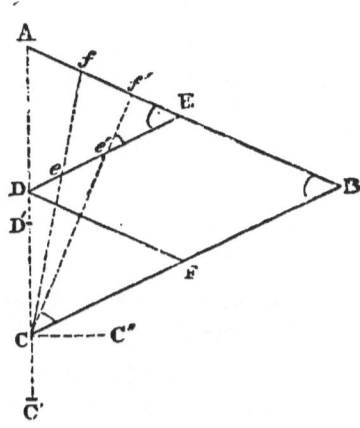

figure to be jointed so as to move universally on a plane. It is clear that if we fix A, and move C to C' in direct line, D would move half this distance and still bisect the line from its new position. Or if we moved C to C" at right angles to the direct radial line from the point A, D would move half the quantity, and at the same angle. Now, by the law, that, if angles are equal and the sides proportional, the whole figure is proportional, every movement of C will give proportional movement to D, which, in this case, will be half. We may also observe that the triangle A B C is proportional in every way to A E D; the internal angle E being equal to B, and the side equal half the entire rule; therefore, any equal angle drawn through the two figures will be proportional, and any angle directed from C will cut equal angles on D E to that on C B. Therefore the inscribed angles and sides terminating in E and B will be equal, and the inscribed areas within the triangle A D E and A B C will be proportional as the squares. By this rule we have only to make any line from C to cut D E proportional to its continuation to the line A E, and the whole movement of the jointed parallelogram or pantagraph will be proportional.

This description may be so far useful, that if we want to reduce in a proportion represented by two lines,—one being not greater than the half of the other,—we may place these two lines together to form one distance, and placing our point of distance on C, open the pantagraph until the full distance falls somewhere on A to E, and the proportion, shown by the quantities which make up the distance, falls upon the line D to E. If we clamp the heads which carry the pencil and fulcrum in this position, the pantagraph will be set to the proportion. In the same manner the dotted line C to f represents reduction to one third, C to f', to a quarter.

Some years since, the writer had his attention particularly directed, by the late Colonel Strange, Inspector of Scientific Instruments for India, to the defects of the old pantagraph, which had remained, in this country at least, without any material improvement for nearly a century. These were, excessive friction, vibration, and contracted action in reductions below one-fifth, caused by the interference of the bars and the weight. These faults are in a degree inherent in the principle of the instrument, but capable of considerable diminution by better construction. The friction occurs principally on the surface upon which the

instrument works, and causes the point not to faith-
fully follow the motions of the tracing point. Indeed,

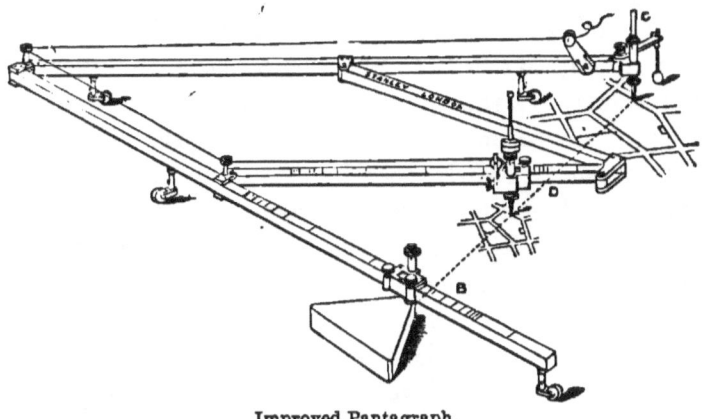

Improved Pantagraph.

however rigid the pantagraph be made, the tracing
point will move a certain distance before the drawing
point starts, which distance will be equal to the sum
of the elasticity of the metal added to the resistance
of the drawing point and the friction of the centres.
For many purposes this small *loss of time*, as it is
technically called, is not of great consequence; where
the pencil line is imperfect, it can be made up afterwards by hand; it, however, makes the instrument
very imperfect and inapplicable for many exact purposes. A partial remedy to the above defect was
discovered by M. Gavard of Paris, and has been adopted
by the writer, which consists in making the pantagraph much lighter, with greater stiffness, by making
the bars tubular; this of course decreases the friction
on the surface of the drawing, and at the same time
prevents the vibration that occurs from the thin flat
arms. The other defect, the contracted action, is

caused by the bars crossing each other and choking the heads, when the pantagraph is being closed, which leaves no room to work, if the reduction be over a sixth, and even at that reduction very imperfectly. The remedy to this is to place both heads, the fulcrum and tracer, on the same side of the bars, instead of placing the second bar between them : the difference of this will be clearly seen by examining the two engravings. A small further improvement in this direction is made by making the weight triangular, instead of circular, which causes it to occupy less of the useable space. There are two other additions in the pantagraph shown on the last page, to which the writer is also indebted to M. Gavard, as well as for the idea of tubular bars : these are a point director and a better plan of holding the check line. The writer would have described M. Gavard's pantagraph entirely, only that, in his opinion, it has one great defect : the centres of the parallelogram are external, so that the instrument will not work near the edge of any board or table on which it may be placed, which must be practically a great inconvenience.*

The EIDOGRAPH, of which an illustration follows, was invented by Professor Willis in 1821. It is a most ingenious and exact instrument, for many

* The author being very busy and unwell when the third edition of this book was required, and wishing to describe this new pantagraph, trusted to a draughtsman on wood to make the drawing from its appearance, set in use, without giving the necessary instruction as to point of view. By this accident the engraving was made rather unintelligible by placing the tracing point C, which is moved by the hand, in the distant perspective. The drawing however is otherwise correct, and as the printer was waiting there was no time to alter it.

purposes superior to the pantagraph, within the range
of its working powers, which however may be considered to be limited to reducing or copying off, between
the full size of the original and one-third the size; for
greater reductions, the balance of the various parts is
thrown so far out that it appears clumsy to use, and is
really inferior to the pantagraph. The great merit of
the eidograph is, that within its range it reduces conveniently and exactly in all proportions; for instance,
we may reduce in the proportion of 9 to 25 as readily
as 1 to 2. It is also in every way superior to the

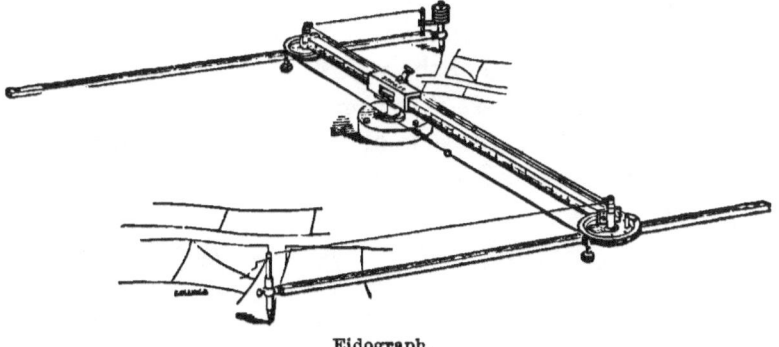

Eidograph.

pantagraph in freedom of action, there being no sen-
sible friction on the single fulcrum of support, and in
its movement it covers a greater surface of reduction.

It is somewhat curious that an instrument of such
great merit should be little known in the profession,
where its uses would be so constantly convenient.
This may partly be attributed to the very few published descriptions, none of which are to be found in
works treating on mathematical instruments. It is
not intended, however, to infer that there are not many
eidographs in use, but that the writer presumes they

are comparatively little known, from his personal
acquaintance with professional men and from the
number of large pantagraphs that are made and sold
to perform work that could be done so much more
exactly and conveniently by the eidograph. This
remark will not apply to the small pantagraph, which
is less expensive than a small eidograph, and answers
perfectly for the reduction of small plans—as, for instance, those frequently attached to leases and con-
veyances.

The details of the construction of the eidograph
are as follows. The point of support is a heavy, solid,
leaden weight, which is entirely covered with brass;
from the under side of the weight three or four needle
points project, to keep it in firm contact with the drawing. Upon the upper side of the weight a pin, termed
a fulcrum, is erected, upon which the whole instrument
moves. A socket is ground accurately to fit the fulcrum, and attached to a sliding box, which fits and
slides upon the centre beam of the instrument. The
sliding box may be clamped to any part of the beam by
a clamping screw attached. Under the ends of the
beam are placed a pair of pulley wheels, which should
be of exactly equal diameters; the centre pins of these
revolve in deep socket fittings upon the ends of the
beam. The action of the two wheels is so connected
as to give them exact and simultaneous motion. This
is effected by means of two steel bands, which are
attached to the wheels. The bands have screw adjustment to shorten or lengthen them, or to bring them to
any degree of tension. Upon the under side of each
of the pulley wheels is fixed a box, through which one
of the arms of the instrument slides, and is clamped

where required. At the end of one of the arms a
socket is fixed to carry a tracing point; at the end of
the other arm a similar socket is fixed for a pencil.
The pencil socket may be raised by a lever attached to
a cord, which passes over the centres of the instrument
to the tracing point. The two arms and beam are
generally made of square brass tubes, and are divided
exactly alike into 200 equal parts, which are figured so
as to read 100 each way from the centre, or by the
vernier cut in the boxes through which the arms and
beam slide, they may be read to $\frac{1}{1000}$ of its half length.

There is a loose leaden weight which fits upon any
part of the centre beam, packed in the box with the
instrument. The weight is used to keep the instrument in pleasant balance when it is set to proportions
which would otherwise tend to overbalance the fulcrum
weight.

In the above details it will be particularly observed
that the pulley wheels must be of exactly equal diameters. It is upon this that chiefly depends the accuracy of the instrument; the periphery of these wheels
being the equivalent to the parallelogram, which has
been already described as the essential feature of the
pantagraph. The adjustment of the wheels to size, by
turning in the lathe, is, perhaps, the reason the results
of the eidograph are more exact than those of the
pantagraph, which has no equivalent compensation
for the always possible inaccuracy of workmanship.

From the details just given, the general principle of
the eidograph may be easily comprehended. Thus, the
wheels at each end of the beam being of equal size, and
the steel bands connecting them being adjustable so as
to bring the wheels into any required relative position,

it follows, that if the arms fixed to the wheels be brought into exact parallelism, they will remain parallel through all the evolutions or movements of the wheels upon their centres; consequently, if the ends of the arms be set at similar distances from the centres of the wheels, any motion or figure traced by the end of one arm will be communicated to the end of the other, provided the fulcrum of support be placed also at similar distance from the centre of one of the wheels.

To adjust or ascertain if the eidograph is in adjustment is very simple, from the reason that when the arms are parallel the adjustment is perfect for all proportions. The manner of ascertaining this is as follows. Place all the verniers at 0, which will bring them to the exact centres of the arms and the beam; place the arms at about right-angles with the beam, then make a mark simultaneously with the tracer and pencil point; turn the instrument round upon its fulcrum, so that the pencil point be brought into the mark made by the tracer; then, if the tracer fall into the mark made by the pencil, the instrument is in adjustment. If it should not fall into the same mark, the difference should be bisected, and the adjusting screws on the bands should be moved until the tracer fall exactly into the bisection, which will be perfect adjustment.

When the eidograph is in adjustment, if the three verniers be set to the same reading on any part of their scale, the pencil point, fulcrum, and tracer will be in a true line. If it should not be so, it would show the dividing or centering of the instrument to be inaccurate. Thus we have a simple way of testing the accuracy of the eidograph in every important particular.

The divisions upon the eidograph do not positively indicate the reductions required to be performed by the instrument, but merely give a scale, which, with the assistance of the vernier, divides the beam and arms into 1000 parts. To obtain the quantity to which the verniers are to be set, it is necessary either to apply to a table of proportions relative to divisions, or by simple arithmetic, as will be shown. A printed table is very generally placed inside the lid of the box in which the instrument is packed, which contains part of the following proportions :—

Table for Reducing or Enlarging Proportions.

Proportions.	Divisions on bars.	Proportions.	Divisions on bars.
As 1 is to 2	... 33·333	As 2 is to 3	... 20
„ 1 „ 3	... 50	„ 2 „ 5	... 42·857
„ 1 „ 4	... 60	„ 3 „ 4	... 14·285
„ 1 „ 5	... 66·666	„ 3 „ 5	... 25
„ 1 „ 6	... 71·428	„ 4 „ 5	... 11·111
„ 1 „ 7	... 75	„ 5 „ 6	... 9·09
„ 1 „ 8	... 77·777		
„ 1 „ 9	... 80		
„ 1 „ 10	... 81·818		

The table here given answers for the general purposes of reducing; such as the bringing of a plan from one chain scale to another, the quantities of which are found by the following rule :—

To find the quantity equal to any given proportion for the setting of the eidograph.—Subtract one sum of the proportion from the other, and multiply this difference by 100 for a dividend.—Add the two sums of the proportion together for a divisor.—The quotient from the working of this will give the number to which the arms and beam are to be set.

For instance, let it be required to reduce a drawing in the proportion of 3 to 5.

$$5 - 3 = 2$$
$$\times 100$$
$$\overline{\hphantom{5 + 3 = 8)\ 200\ (25}}$$
$$5 + 3 = 8)\ 200\ (25$$

The centre beam is to be set to 25 on the side nearest the pencil point, the pencil arm is also set to the 25 nearest the pencil point, and the tracer arm is set to the 25 farthest from the tracer. If it were required to enlarge in the same proportion, each side would have to be set at the opposite 25.

To clearly illustrate the subject, it may be well to give another example. Let it be required to reduce an ordnance plan of five feet to the mile to a scale of three chains to the inch. First, we must have like terms, therefore to reduce both proportions to feet to the inch will, in this instance, be the most simple way; thus :—

5 feet to the mile = 88 feet to the inch.
3 chains to the inch = 198 feet to the inch.
$$198 - 88 = 110$$
$$\times\ 100$$
$$\overline{\hphantom{198 + 88 = 286)\ 11,000\ (38.461}}$$
$$198 + 88 = 286)\ 11,000\ (38{\cdot}461$$

If the slides of the instrument be set to 38·46, it will be, practically, sufficiently near.

Since the last edition of this work, in which the above description was given, the writer, in endeavouring to correct the defects of the eidograph has been able to make some important improvements. It has not been thought well, however, to disturb the description given, as the greater number of instruments in use will be for a long time to come of the old construction.

The illustration below shows the new instrument. The improvements are principally additions, and therefore it would not be difficult to have them adapted to any instrument of the old construction. As stated, the eidograph would not properly reduce below one-third. In its improved form, to be described, it will reduce in all proportions to one-eighth or even fairly to one-ninth. The changes and additions to effect this, consist of supporting the main beam instead of balancing it, when

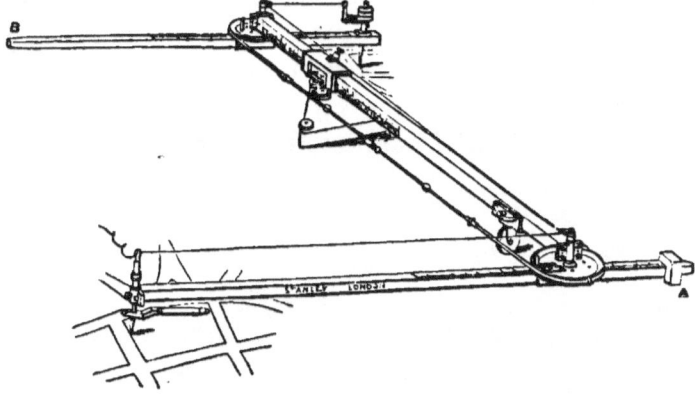

Improved Eidograph.

using the instrument for reductions below one-third, by means of a moveable castor, the roller of which is two inches diameter, fixed on the beam with its axis in a direct line to the principal fulcrum, on which the instrument moves; this causes very little friction, and dispenses with the balance weight on the beam, at the same time giving two points of support. Another improvement is that of making the tracing and drawing point exchangeable. When this is done, one arm of the eidograph may be reduced to about half of its length, and the lopsided strain which the ordinary eidograph presents in reductions of a third, will be materially modified. A further improvement in the now short arm is made by fixing the pencil-holder at about two inches from the end of it, and placing a small separate balance weight on it when it is required. The weight will be placed behind the end axis for small reductions, between the axis and tracer for from fourth to sixth, and beyond the tracer for extreme reductions, seventh to ninth; it has a table engraved upon it for its position for each reduction. Another small improvement is that of making the fulcrum weight triangular instead of circular, which form encumbers the drawing less. There are some other small details of improvement of less importance.

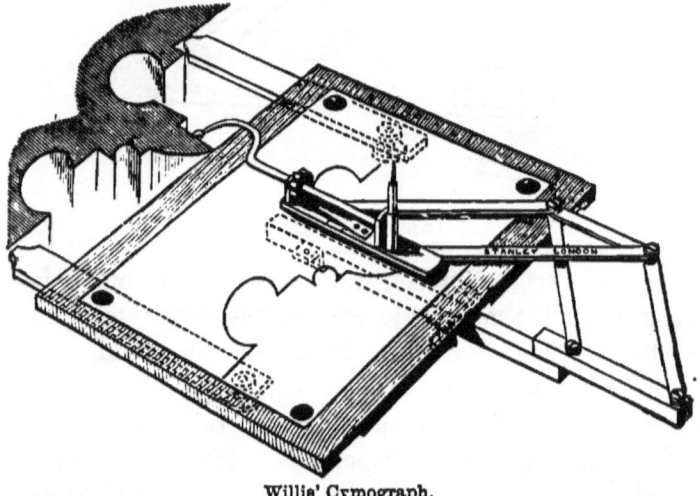

Willis' Cymograph.

The CYMOGRAPH was invented by Professor Willis, in 1838, and has been somewhat improved in detail since. It was not described in previous editions of this work, the writer·thinking it more ingenious than useful; but as he has found the instrument brought forward again, and has met with a few persons who have found it available for certain purposes, he now introduces it, though still with not much idea of its practical value. It may be used for copying full size, but better for taking the outlines of solid objects, as mouldings, carvings, etc., or taking off outlines from drawings to material for pattern-making or carving. The instrument consists of two free-jointed parallelograms of metal; one of these, which is made of shorter sides than the other, is fixed to a small drawing-board by a moveable stay projecting at right angles to the edge. The parallelogram with long sides is jointed to the first, leaving one end free to carry the tracing bed, which is constructed as follows. A straight rod is placed in bearings in a line with the end of the parallelogram, so that it can rotate on its axis. The rod is bowed at the free end, but its point is brought back to the axial line, where a small knob is placed to form the tracing or following point. The bow above described, which will stand in any direction to the axis, is to allow the tracer to follow any undercut projection in the moulding or figure to be traced. The copying pencil point is placed on the bed in a line with the axis of the tracer. The writer has introduced two slides to the board, which move out from the edge and fix when required, to enable it to be held steadily against any irregular figure; even with these, the instrument is a little troublesome to use.

K

CHAPTER XIX.

INSTRUMENTS INTENDED TO FACILITATE THE DELINEA-
TION OF NATURAL OBJECTS, BUILDINGS, ETC.—CAMERA
LUCIDA—AMICI'S CAMERA—OPTICAL COMPASSES, ETC.

MANY instruments have been invented to simplify the
art of sketching natural objects and art forms; truly,
they can scarcely be said to offer other than false helps.
We have no instrument that is capable of producing
equivalent artistic results to those that may be attained
by studious observations, united with moderate practice. To the young student in sketching, the writer
would say—Persevere with the pencil only; observe
truly, work patiently, and avoid all helps from instruments; for these, being inartistic aids, will leave by
their attained use neither the pleasure nor satisfaction
that will naturally be experienced by every one who
may feel himself in some degree a true artist. Further, we cannot have instruments always at hand, and
the attainment of an easy power of delineation is not
an indifferent matter to any practical man—profession-
ally it is imperative. Many of our great men have
seemed to speak with the pencil, and thereby illustrate
ideas quicker and better than words could express them.

If any instrument be employed for sketching, except
so far as some trifling aids to be mentioned further on,
the camera lucida is undoubtedly the best, for this apparently lays the complete reduced image upon the
paper in all its natural colours and shades; it further
only requires a steady hand, and some practice in the

management of the instrument, to trace the forms, and produce an accurate outline. Thus it may in some way be made to supply the deficiency of art education, or the want of natural power of imitating form.

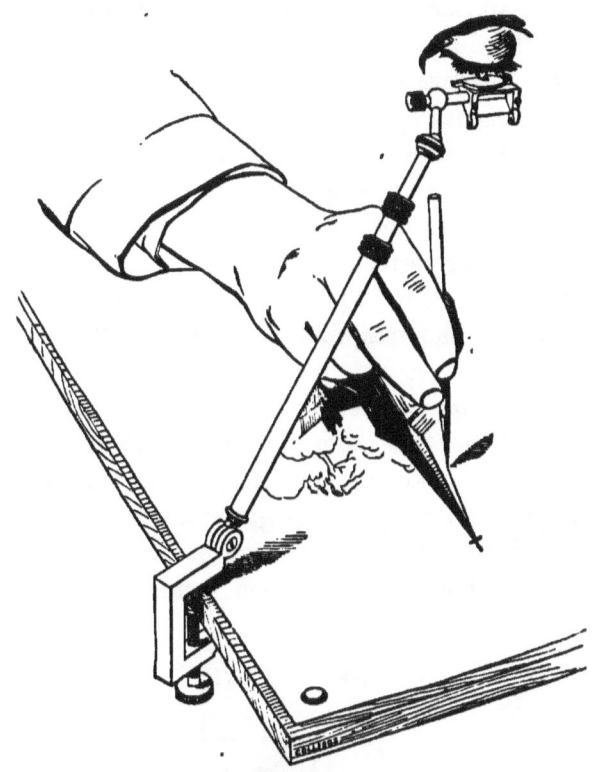

Wollaston's Prismatic Camera Lucida in use.

The CAMERA LUCIDA was invented and patented by Dr. Wollaston in 1806.* In its most simple form it may easily be constructed experimentally. If a small piece of tinted plain glass be fixed by any simple contrivance at the angle of 45 degrees over a sheet of paper,

* Specification No. 2993. 1806.

say at one foot distance from it, and the eye be brought over the glass so as to look through it to the paper, any object in front of the glass upon which there is a strong light will be partially reflected to the eye, and a faint image will appear superimposed upon the sheet of paper. If a pencil be brought upon the paper, it will be seen through the glass sufficiently to trace the outlines of the forms as they appear. To enable the eye to return to the same position, a piece of black card with a hole through it may be fixed above the glass. By this simple camera lucida the image appears inverted, therefore it does not answer perfectly for sketching; it is, however, much used for scientific purposes, and answers perfectly for sketching the magnified image projected by the microscope.

The camera lucida used for art purposes is constructed in such a manner that the image shall be twice reflected, so as to appear in an erect position. This is effected by a solid quadrilateral prism of glass, the cross section of which is shown by the illustration below. The vertical and horizontal sides of the prism form with each other

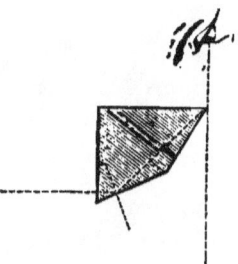

Section of Wollaston's Camera Lucida Prism.

an exact angle of 90 degrees, each of the other sides being inclined to these at the angle of 67½ degrees, and forming at their meeting the angle of 135 degrees; this last is termed the reflecting angle of the prism.

By this construction of prism, according to the law known in optics as prismatic reflection, the rays of light reflected from an object will pass in direct line through the vertical and horizontal surfaces, and be reflected by the inclined surfaces. The direction the rays will take is shown by the dotted lines in the engraving, by which it will be seen that the rays are twice reflected—that is, once on each of the inclined surfaces; therefore the image will appear erect to the observer.

It is not possible to see the pencil through the prism, as it is seen when the plain glass is employed, as the object is *completely* reflected, and not *partially* so, as in the simple camera lucida; therefore the pupil of the eye has to be brought to observe the pencil by direct vision over the edge of the prism, which requires some practice to effect with comfort.

The prism being a perfect reflector, too much light from the object will often proceed to the eye to enable the draughtsman to observe the pencil simultaneously with the reflected image, unless the upper surface of

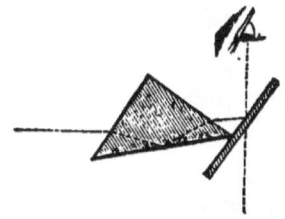

Section of Amici's Camera Lucida Prism.

the prism be nearly covered; for this reason a shutter, with a small hole through it, is placed over the prism.

The shutter is movable horizontally, so as to cut off the light until the eye receives a faint image only, which enables the pencil only to be seen sufficiently to trace the reflected object.

AMICI'S CAMERA LUCIDA is a modification of that described by Wollaston. It is preferable in some respects, particularly in requiring less practice to use efficiently, and in admitting the eye to change its direction within certain limits. It may be considered as a combination of Wollaston's simple and prismatic camera lucidas; a prism being used to erect the image only, and a piece of parallel glass at 45 degrees being employed as in the simple camera lucida to partially reflect it. The last engraving gives the form and position of the prism and parallel glass, the dotted lines showing the direction which the reflected image takes so as to appear superimposed upon the drawing surface. With this camera lucida the parallel glass will generally reflect too much light from the image to the tracing pencil to be clearly seen. As the complete instrument is generally constructed, it admits of one or two moveable shutters of tinted glass being placed before the prism, to shut out the surplus light reflected from the object to be drawn.

The prisms of either of the above-described camera lucidas having all flat surfaces, will be in focus at any distance above the plane of delineation, the reflection being enlarged in proportion as the prism recedes from the plane. For the same reason no image will appear distorted by the reflection, as it would if it passed through the curved surfaces of a lens, as may be observed in the camera obscura.

The illustration on page 131 represents the instrument complete. The optical portion of either kind of camera lucida described is enclosed, except the necessary aperture, in a light metal box, which is attached by a universal joint to the top of a sliding tubular stand, which is again jointed at its lower end, so that the instrument may be adjusted to any position above the drawing-board or table, to which it is attached by a kind of cramp.

The camera lucida may be used for copying drawings to reduced size, or small objects placed in favourable light and position. In all instances the board to work upon must be fixed upon a stand or firm table, as it must not be moved by any chance during the taking of one picture, or it would be reset with great difficulty.

OPTICAL COMPASSES.—The instrument illustrated upon the next page may be used to obtain the exact relative perspective position of two objects; and may be used for assisting in sketching general views, buildings, astronomical observations, etc. In optical principles this instrument is similar to the sextant— that is, it obtains apparent coincidence of a reflected with a direct observation. It is different from the sextant only in that the observation, instead of being read upon a divided arc, is taken by the points of a pair of compasses, so that it may be at once transferred to a drawing.

The following details of construction will render the principle clear to any one unacquainted with the sextant. The mechanical portion of the optical compasses consists of a pair of ordinary drawing compasses, the legs of which are made tubular, so that the points may draw out some distance. By this arrangement the drawing may be varied in size without the draughts-

man changing the distance of his position in relation
the object.

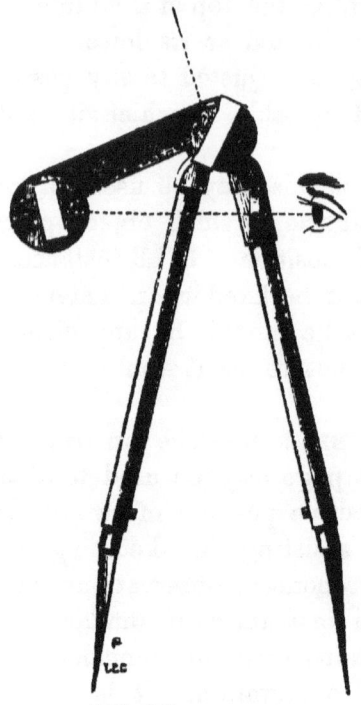

Optical Compasses.

The optical portion of the instrument consists of a
mirror fixed vertically to one leg of the compasses,
directly over the head-joint, so that it follows the angle
of the opening of the compasses. Projecting from the
other leg of the compasses, from near the head-joint,
an arm is carried out a short distance, and upon the
outer end of this arm another mirror is fixed to face
the first. The lower half only of this mirror is silvered,
the other half being plain glass. Upon the same leg as
that on which the arm is attached, at a short distance

from the joint, is fixed a piece of metal with a small hole through it, which forms the eye-piece. Any object in front of the mirror upon the joint will be reflected to the mirror upon the arm, and its image may be clearly seen in this mirror by looking through the eye-piece; at the same time the observer may see an object through the plain part of the glass by direct vision. Thus two objects will appear, one over the other, the opening of the compasses at the time being the distance of these two objects, according with the perspective scale to which the compasses are first set. The direction of the reflected and the direct vision is shown by dotted lines upon the illustration.

By this instrument, as with the sextant, angles of position of objects to the observer are taken up to 120 degrees, although the compasses open only to 60 degrees. Of course this is immaterial to a drawing instrument, as the proportions are relative to each other, and correct to perspective scale, which is all that is required, provided the compasses produce a drawing of sufficient size. If the compasses are required to take observations for a large drawing, it will generally be necessary to double the distance of the opening, which is easily done by rolling them over on one point. Although these compasses may be used in any direction, it is generally most convenient to use them horizontally or vertically.

The method of using the instrument to obtain the perspective distance of two objects on the same horizontal plane, would be to hold the instrument horizontally in the right hand by the small handle at the back of the head-joint, and to look through the eye-piece and the plain glass at the object nearest the right hand; then, if the compasses be gradually opened with the left hand, the second object will appear in the mirror under the first. The distance of the opening of the compasses may then be transferred to the drawing.

If the drawing is intended to be made of a given size, an observation should be taken with the instrument of the extreme of the object, and the legs of the compasses should be drawn out until they reach the required size, or half the required size if the drawing be large, and it is intended to double the distances of the opening of the points.

The reader is aware that the author does not recommend instruments of any kind for sketching. There may be cases in which the above-described instrument would be useful, as, for instance, in sketching a difficult perspective view of a street, a few principal objects might be taken and placed in true position and size, which would serve as a guide to the whole perspective. It might also be conveniently used where the whole perspective cannot be taken at one observation, which frequently occurs, as in sketching a cathedral interior or exterior, where it is necessary to be close, and very difficult to judge of comparative perspective heights. If a few altitudes be taken with this instrument, the details would be filled in with greater certainty.

The following simple contrivances, which cost only a trifle, may be found useful to many, as the most of us experience some difficulty in sketching from nature. Thus, one person who may be good at details and rounded forms or colour, may fail at perspective angles; or, if efficient at observation of these, may fail at proportional equal reductions, or artistic effects, or otherwise, that the following little instruments may possibly obviate.

PERSPECTIVE DIRECTOR.—A very simple little instrument that will be found of great use to such as have difficulty with perspective angles. The instrument need not be larger than a pocket knife, say four inches by half an inch by a quarter inch. It consists

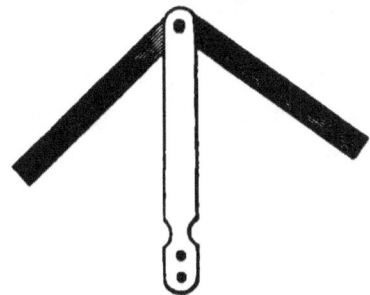

Perspective Director.

of two thin blades of ivory, which close into a handle of equal width. The handle forms the vertical of any line or building, and the two arms will subtend any angle to this. If the instrument be held up in front of the observer before any building or other object, the blades may be adjusted to follow the perspective lines by looking over them to the lines required, keeping the handle erect by looking in the same manner at any wall or other vertical line. This angle may be laid upon the drawing direct and marked off correctly.

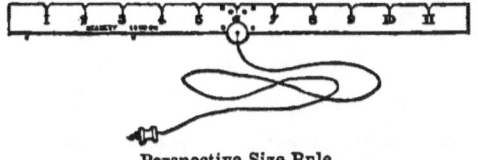

Perspective Size Rule.

PERSPECTIVE SIZE RULE.—Some persons experience great difficulties in relative sizes of parts of a building, landscape, or other object, being unable to get all parts proportional upon a reduced scale. To such the writer's perspective size rule will be of immense service. Indeed, to such as do not feel a difficulty, it may be useful to arrange a picture for the best effects. This instrument is represented in the engraving. It is simply a twelve inch flat rule, with a joint in the centre to close to six inches. This length is convenient for general purposes, but the rule might be made shorter or longer if the scale of the drawing renders this advantageous. In the engraving it is divided into twelve inches, and a large distinct figure is placed under each division. Deep notches are cut in one edge to the divisions. A small hole is made through the centre of the joint, a piece of Indian twist is threaded through this, and a knot tied to fix the twist from slipping through the hole. The twist is left of a convenient length, according to scale required, from fourteen to twenty-four inches. A small ivory reel is placed on the twist, something like a small shirt stud, to hold between the teeth; this has a hole through it, similar to the joint of the rule, to fix the twist by a knot.

To use the perspective size rule.—Having adjusted the string to length required for scale of drawing, the button is placed in between the front teeth, and the rule is held out the length of the string. Now by looking over the notches by one eye at any portion of the building, object, or landscape, the part can be measured off from point to point, and transferred direct to the drawing. The rule may be used in any direction, and horizontal, vertical, or perspective measurements be made. It will for most draughtsmen only be necessary to make a few dots for general distances, the interspaces being filled in by the eye. The joint of the rule will give a perspective angle also, if desired.

Perspective Window.

PERSPECTIVE WINDOW.—Having been in the habit of sketching for some years, on my annual holiday tour, I have found the following little contrivance very useful to apply to the prospect or objects before commencing the drawing, to get the most artistic picture. Having my sketching block that I intend to use for the sketches, I take a card and cut out the centre part to a scale that will be exactly proportioned to my block. Say I am going to use a block 10 × 8 inches, I take a card of about this size and cut an opening in the centre of it, say two-thirds the size, that is 6⅔ × 5⅓ inches. This I hold up before my eye upon the field of view at different distances, vertically or horizontally, and observe the picture that appears in the opening. By this means, shifting about a little, the best effect is soon found, with the amount of agreeable foreground and sky, or portion or position of building to be sketched, and the picture is sure afterwards to please. I have two ink lines on the card, which give me centres, which are generally all the measurements that I care for. The *perspective window* is carried conveniently in the pocket of the sketch block, and will do for any future block of the same size or proportion.

There is yet another method of sketching, just perhaps worth mentioning to make the subject complete, which appears to answer, sometimes practised. This is by having a large convex mirror placed in front of the object, the mirror being blackened instead of silvered, to subdue the light. The object in this case appears as a picture of the size to be copied. I have seen this apparatus used abroad, but not in this country; it is scarcely an artistic means.

Binko's Spectrograph, a sketching instrument which has been lately introduced as a kind of toy, gives a reflected image on the principle of Pepper's ghost; and as it reverses the image right to left, it may at some future time be constructed into a useful tool for

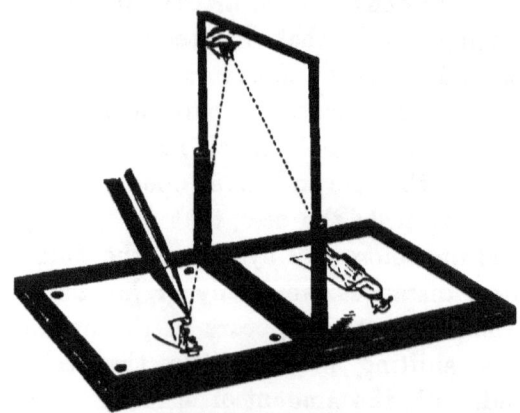

Binko's Spectrograph.

the lithographer or wood-engraver. The instrument consists of a board, upon which is fixed, vertically across the centre, a piece of plain glass. If a drawing be placed on the half of the board next the light, and the eye be on the same side of the glass, a reflection will appear on a piece of paper placed on the other half of the board, that can be seen through the glass sufficiently well to trace the image, which is, as stated, reversed.

Before leaving this elaborate and in some respects not very practical subject, it may be well just to mention other plans which have been invented with the good intention of rendering sketching easy, such as—several modifications of the pantagraph, the tracer of which follows the apparent lines upon a sheet of glass; instruments with wires or threads over frames placed before the observers, either to observe distances or to divide the picture into squares; and a few other schemes, which are generally really of as little practical value as they are in principle antagonistic to true art.

SECTION II.

Relates to Drawing Instruments used as Guiding Edges, Instruments for Measuring, and Drawing Materials.

CHAPTER XX.

SURFACES TO DRAW UPON—DRAWING-BOARDS—TRACING FRAMES—PLANE TABLES—SKETCHING-BOARDS AND BLOCKS—ENGRAVER'S TRAY, TRESTLES, ETC.

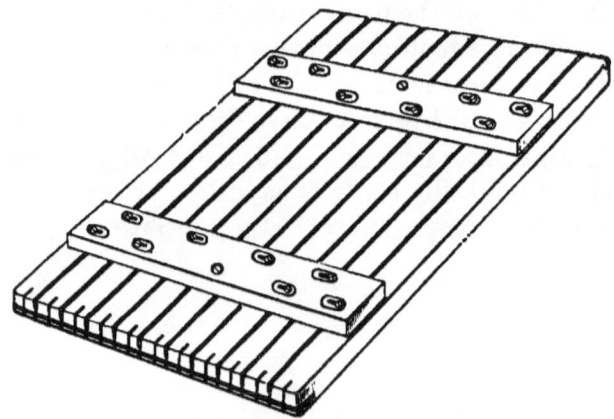

Improved Drawing-board.

A GOOD DRAWING-BOARD is a very great desideratum to the draughtsman. The qualities it is important that it should possess are—an equal surface, which should be slightly rounded from the edges to the centre, in order

that the drawing-paper when stretched upon it may present a solid surface; and that the edges should be perfectly straight, and at right angles to each other.

These qualities seem theoretically easy to obtain in a material so tractable as soft pine-wood, of which drawing-boards are generally made. Practically, this is very difficult, as wood, however well seasoned, is continually changing its form, rapidly absorbing moisture from the atmosphere, causing expansion of the fibre, and slowly contracting unequally as the moisture evaporates, and this with a force no simple means will resist.

For these reasons, the true principle of making a drawing-board is that which will leave the wood free, so as to allow these changes to take place without materially affecting the surface or square of the board. This is nearly effected in a drawing-board invented by the late elder Mr. Brunel, of which the back view is shown in illustration. The front surface is quite plain. The construction is as follows. The board is glued up to the required width, with the heart side of each piece of wood to the surface. A pair of dry hardwood ledges are screwed to the back side. These screws pass through the ledges in oblong slots, bushed with brass, which fit closely under the heads, and yet allow the screws to move freely when drawn by the contraction of the board. To give the ledges power to resist the tendency of the surface to warp, a series of grooves are sunk in half the thickness of the board over the entire back. These grooves take the transverse strength out of the wood, and allow it to be controlled by the ledges, leaving at the same time the longitudinal strength of the wood nearly unimpaired.

L

It may be observed that this board has two working edges that present the end of the grain, which is unpleasant to work against with the square. To obviate this, a slip of hard wood is let into the end of the board. The slip is afterwards sawn apart at about every inch, to admit of contraction.

Drawing-boards ledged on the above principle may be made without the grooves in the back surface, but the board to *stand* must be made much thinner in relation to the ledges. This construction is not so good in any way as that first described.

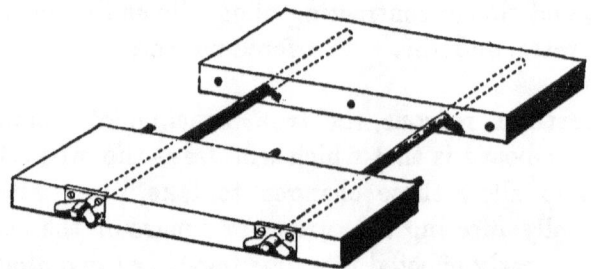

Perspective view (centre part removed) of Patent Portable Board.

The writer has lately patented a new plan of making drawing-boards, which appears to answer perfectly, at the same time admitting them to be made portable. The construction is as follows. The board is made up in three or four widths which are dowelled together. Two tempered steel bolts are passed edgeways through these at the places usually occupied by the ledges. The separate pieces are grooved as those first described, but the grooving is a saw-cut only, and not carried quite to the ends. The separate parts can be detached from the bolts for packing. The advantage of these boards for hot climates is, that they may be screwed up by the bolts as the wood shrinks; the saw-cuts at the back also allow the board to be pulled slightly rounding, to make the surface solid to draw on. They are not recommended for home use, as the boards described at the commencement of the chapter answer satisfactorily, but for sending abroad they will often cost less than any other kind. A single board may also be parted and strapped together or put into a solid leather case for travelling, where the ordinary board would be quite unavailable.

If these boards for tropical climates are not required to be made portable, bars of steel are let into the ends, and answer the same purpose as the bolts, and are somewhat stiffer.

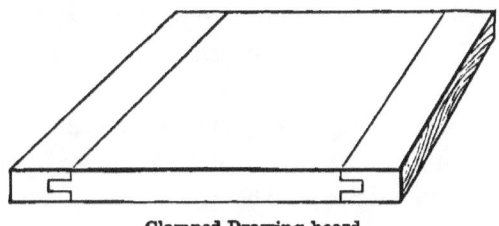

Clamped Drawing-board.

The CLAMPED DRAWING-BOARD, the perspective view of which is shown in illustration above, is of a very general but most defective construction. It is an instance of the attempt to make one piece of wood resist the contraction and expansion of another, the effect of which is that the board generally warps and splits, and is never square; it may, however, be used as a sketching-board, if made under sixteen inches wide.

The PANELLED DRAWING-BOARD, of which the next illustration is a half perspective transverse section of the back, consists of a frame with an internal rabbet, into which a loose panel is fitted with a corresponding rabbeted edge, so that its surface comes even with the frame. The panel is held in position from the back by two loose ledges, the ends of which fit into the frame. This board is useful for sketching purposes, as a damp

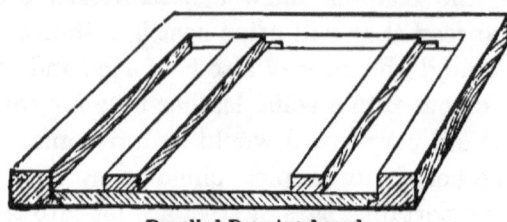

Panelled Drawing-board.

sheet of drawing-paper pressed in with the panel becomes perfectly stretched when dry. It is totally unfit for use with the tee-square, from the reason that the pressure of the paper causes the sides of the frame to bulge out.

Another sketching-board is made similar to a plain clamped board, with the addition of a row of projecting pins round the edges of the top surface, and a light open frame hinged to it, so as to fall over and protect the pins, the frame being held down on the side opposite the hinges with a pair of hooks and eyes. Upon this board six or eight sheets of paper may be damped and laid over the pins, and will all be stretched when dry, so that each may be removed separately when the drawing is completed, leaving the next surface ready to be drawn upon. The frame being raised above the surface of the paper, makes this board difficult to be used with the tee-square.

For drawing upon small wood blocks for the engraver, also upon lithographic stones, the writer has constructed a very convenient drawing-board in the form of a kind of tray, with a rim an inch thick firmly secured. The rim rises from the inside surface of the bottom of the tray, nearly the thickness of the wood block, which is held firmly in its required position by a pair of wedges that clamp up quite square, and

Wood Draughtsman's Tray; also Centrolinead, described at p. 169.

parallel to the inner edge of the rim. The outer edges of the board or tray give direction to the tee-square.

SKETCHING-BLOCKS answer the purpose of sketching-boards, and have justly superseded them, particularly for small sizes. They consist of many sheets of drawing-paper, united by the edges, which are planed quite square; the whole forms a solid block of paper of about half an inch in thickness. When a drawing is completed on the upper sheet, this may be lifted off by passing a penknife between it and the next sheet, which will then leave a surface ready to receive another drawing. Sketching-blocks are sometimes placed in covers as a portfolio, and are then termed *block-books*.

A very convenient kind of drawing-board for copying drawings is that termed a Tracing-frame, the peculiar merit of which is that it is a means of copying drawings without the slightest soil or injury. It consists of an ordinary open square frame of stout wood, into which a sheet of plate-glass is sunk to the level of the surface.

A second similar frame, but without glass, is hinged to the first, so as to take a position immediately under it, and to form a bed to rest on any flat surface. A pair of struts with rack notches upon this, support the upper or tracing frame at any angle convenient to receive the necessary light.

To use the tracing-frame, the original drawing is first placed over the glass, and a plain sheet of paper pinned or otherwise fixed over it. The frame is then placed at an angle to the light, so as to make the whole appear transparent. The lines of the copy may now be traced conveniently and with great accuracy. This plan is much resorted to for the preparation of the drawings on parchment, accompanying specifications for letters patent. The only inconvenience in the above method is that the drawing has to be placed slantwise to transmit the light, which renders the frame somewhat awkward to work upon.

The writer has constructed a Copying-table which admits of the surface being left level, the light being obtained through the surface glass by reflection from a mirror, placed at an angle of 45 degrees in a frame supported by the legs of the table. To use this table with advantage, it must be placed in front of a very low window, or be raised to the height of the window. If used in a dark neighbourhood, a second reflector will be required outside the window. If too much light falls on the surface of the copy, it requires shading with a holland blind. Where much copying is done, it will be worth the expense of having the building adapted to this method of copying, which is the most expeditious and exact.

Copying-table.

Drawing-boards of every description are made to set sizes only, according to the papers intended to be used upon them, a little extra being allowed for margin; they also take the names of the papers for which they are made. The sizes almost universally used are—antiquarian, 55 by 33 inches; double-elephant, 42 by 29 inches; imperial, $31\frac{1}{2}$ by $23\frac{1}{2}$ inches; and half-imperial.

The PLANE TABLE is used for small surveys of details of land which are made on the spot. It consists of a small drawing-board that can be fixed upon a firm tripod-stand, and adjusted horizontally. A loose moveable rule, upon which is fixed a pair of direction sights, forms a part of the instrument, and is used upon the table to take the direction of any object which, being sighted, will correspond with the ruling edge of the rule. A magnetic compass is attached to the rule, so as to give the direction of its position when set for use. The distance of the object is measured off upon the land by tape or chain, and plotted on the table by scale. It will be observed that the plane table works all the actual visual angles that are tied by scale from measurement, which enables the complete plan to be made at once. The plane table is much used in the army and in India, and but very little in this country. It is sometimes made with a telescope attached to the rule, with cross webs in the optical axis to take bearings exactly.

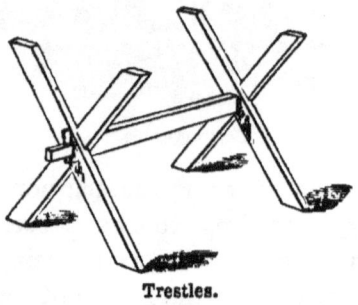

Trestles.

Drawing-boards of large size are generally supported upon Trestles, the most convenient form of which is shown in the illustration above, which consist of two crosses of stout, hard wood, which are halved together; a square mortise is made through the intersection, through which a shouldered tenon on the longitudinal bar passes a sufficient distance to receive a mortise and key-wedge, which forces the cross against the shoulder of the bar, and at the same time fastens the whole to-gether. When the wedges are tightened, these tres-tles are very firm. They are also quite portable when the wedges are withdrawn, every piece being thereby separated.

CHAPTER XXI.

Straight-edge.

THE STRAIGHT-EDGE is a flat blade of wood or metal,
and is used for guiding the pen or pencil to produce
straight lines. It has generally one of its edges bevelled,
which is the one used for drawing lines; the thickness
of this edge should be about one-fourteenth of an inch,
to be suitable for use with the drawing pen. If the
edge be made thinner, the close contact of the point of
the pen causes the ink to run down upon the drawing.
The above thickness of edge is practically the best for
every description of undivided ruling edge. If straight-
edges be made of wood, they are better made in three
widths glued together, the two edges being cut from
one piece of wood; in this way the warping tendency
of one piece of wood is counteracted by the other. A
mahogany blade with two edges of ebony answers
very well for short lengths. Wooden straight-edges are
efficient for architectural or mechanical drawing and
perspective. In civil engineering for base lines, steel
straight-edges are unquestionably the best; these need
no particular description, as they differ from wooden
ones only in the material, except that they have generally one very thin edge for pencil, and a thicker one
for ink lines.

The writer has patented a new method of making straight-edges which is also applicable to the blades of tee-squares. From experiments only, it appears to

Section of Patent inlaid steel Straight-edge.

answer well. The plan is to make a fine saw-cut from the edge of the straight-edge to nearly its entire width, and to insert in this a piece of thin steel : what is termed busk-steel answers best. The steel is left free in the wood except near the edge by which it enters, which is riveted to it. When the straight-edge is bevelled, the steel stands on the top edge of the bevel to resist the wear. This produces an economical steel straight-edge, as the steel requires no finishing, with the advantage that it does not attract perspiration to soil the drawing or to become rusty. The steel being wide controls the warping tendency of the wood.

There have been many suggestions for testing the accuracy of straight-edges, some method being very necessary, as wooden straight-edges are liable to warp, and steel ones are frequently sold very inaccurate. One of the most simple means, which is effective when very great care is used, is to lay the straight-edge upon a stretched sheet of paper, placing weights upon it to hold it firmly; then to draw a line against the edge with a needle in a holder, or a very fine hard pencil, held constantly vertical or at one angle to the paper, being careful to use as slight pressure as possible. If the straight-edge be then turned over to the reverse side of the line, and a second line be produced in a similar manner to the first, at about the twentieth of an inch distance from it, any inequalities in the edge will appear by the differences of the distances in various parts of the lines, which may be measured by spring dividers, or perhaps quite as accurately by observation.

A' very simple method will be found to answer very well if three straight-edges are at hand; this method is used in making the straight-edge. Two straight-edges are laid together upon a flat surface, and the meeting edges examined to see if they touch in all parts, reversing them in every possible way. If these two appear perfect, a third straight-edge is applied to each of the edges already tested, and if that touch it in all parts the edges are all perfect. It may be observed that the first two examined, although they may touch perfectly, may be regular curves; but if so, the third edge applied will detect the curvature.

A method recommended in a work on mathematical instruments is to hold two edges together up to the light, which they will exclude if they be perfect. As it is impossible for the human hand to hold two thin edges together square to their faces; and even if possible would be no test, as they might be curved, therefore this rule is of little value. It is only mentioned here, as the author has seen persons attempt to avail themselves of it.

In making large drawings it may happen that the straight-edge at command is not of sufficient length to produce the required line, which is frequently the case in laying down a long base line. One method of getting over this difficulty is to piece the line out; that is, to draw a line as long as the straight-edge—afterwards to lay the straight-edge over the line, letting it pass some distance beyond it at one end—and to draw in the projecting piece. The theory of this method is correct, but in practice it is found very defective, except when a very short extra piece is required.

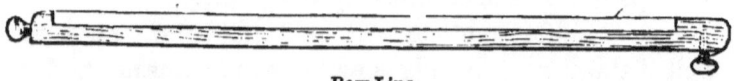

Bow Line.

When a line considerably longer than the straight-edge is required, a bow line will be found a simple and effective contrivance. This is merely a slip of wood which has a small piece of hard wood fastened upon one side of it at each end, across which a fine wire is stretched; a plain key at the back of the lath, similar to the key of a violin, will tighten the wire to any degree. When the bow line is laid over the drawing, it rests a short distance off the paper. To produce a long line with it, it is necessary to make marks exactly under the line at distances apart that a straight-edge will reach; the bow line is then removed, and the straight-edge is used to produce the line. This bow will also form a very good test for a straight-edge.

The line may be produced without the bow by merely stretching a wire or silk cord across the drawing; but this method is less convenient for testing the accuracy of a line after it is produced, as it is difficult to stretch a cord a second time over it, or the paper may touch the cord somewhere and distort it.

Another method of dotting a long line is to stick a needle in at each end of the required line, and get an assistant to move another needle held erect in a holder about the centre of the intervening space, until it appears in a line with the two first, when looking along the surface of the drawing. This is called a picket line, and is similar to the method employed for laying the base lines in a small survey by the chain only.

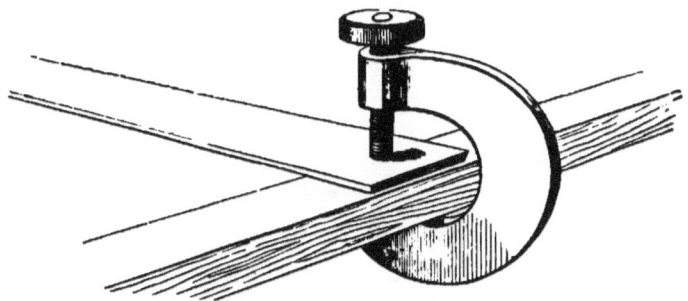

Clamp and Straight-edge.

Occasionally in plotting sections it is found convenient to fix the straight-edge, the scale and offsets being used above it and against it. This is sometimes done by leaden weights. If the straight-edge reach to the edge of the board, it may be perfectly secured by the simple clamp illustrated above. This clamp is also very convenient for holding a steel straight-edge to cut off a drawing from the board, or to hold the straight-edge for cutting card in modelling.

CHAPTER XXII.

Plain Tee-square.

THE T or TEE-SQUARE is used for making horizontal and sometimes vertical lines upon a drawing. It can only be used with a drawing-board, the edges of which direct it, and keep its working edge constantly in one parallel.

The ordinary form of tee-square is represented in the illustration above. It consists of a stock of wood of about three-quarters of an inch in thickness, with a mortise in the centre of one of its edges, into which a blade similar to one of the straight-edges already described is fixed. The edge of the stock which receives the blade is generally rabbeted from each side, so as to leave a tongue of the same thickness as the blade, standing along the centre of the edge. A square is most convenient if made the length of the drawing-board with which it is to be used. The blade should be about one-fourteenth of its length in width, and one-eighth of an inch in thickness. The stock should be about one-third of the length of the blade, and of rather less width. Tee-squares should be made of hard wood; pear-wood answers very well, from the equal density of its texture; or mahogany edged with ebony, in the manner recommended for straight-edges, makes an excellent blade. Squares made entirely of ebony soil the paper, from the oily nature of the wood, which collects dirt that comes off afterwards upon the paper. Squares made of mahogany or satin-wood present naturally a crumbly, soft edge. Steel blades are also very unfit for drawing squares; metal of any description attracts the perspiration of the hand, and soils the drawing by working up the black-lead.

The French glue a tongue of brass in the edge of a pear-wood blade. At first the edge appears very neat, but the contraction and expansion of the metal soon either loosens the edge, or draws the blade out of truth. The writer's method of making blades, described in the last chapter, it is hoped will be better. It is customary to polish squares, but this is better left alone; polishes, varnishes, or oils are very objectionable upon any kind of drawing instrument. For porous woods, as mahogany, a mixture of shellac in absolute alcohol rubbed over the squares will cement up the pores of the wood : this may be entirely cleaned off with glass-paper a few hours after the application, and will leave the surface of the wood smooth and clean.

In using the square, the rabbet is placed over the edge of the board, so as to bring the blade down upon the surface of the paper, the blade being guided by gentle pressure of the centre of the stock against the edge of the board. The tongue along the edge of the stock is intended to keep the rabbet at an equal distance down the edge of the board; otherwise, should the edge not be quite square, the tipping of the stock would throw the blade out of truth.

The Continental manner of making the tee-square is to let the blade into one side of the stock, thus leaving one side only rabbeted and the other flat.

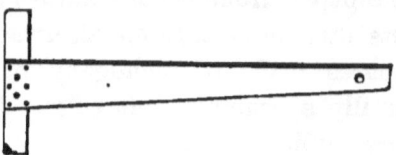

Improved form of Square.

The author's plan of making a square, which has met with almost universal approbation from the profession, is illustrated above. It is of somewhat more simple construction than those already described, being merely a blade of wood screwed upon a stock, without being either sunk or mortised into it.

When in use, this construction of square presents several advantages; one of the most important of which is, that the upper surface of the stock sinks to the level of the surface of the board, and therefore allows the scale or set-square to pass along the blade over the stock, and to produce lines to the edge of the board. Another advantage is, that the blade being screwed upon the stock, it may be taken off at any time to shoot the edge, should it become notched or out of truth The blade is made narrow towards the point, and broad at the stock; this taper form, although light, presents great strength to support the point from deflection, thus obviating a fault common to all ordinary parallel blades. The width of the blade at the stock also does away with the necessity of a rabbet, a convenience which allows the edge of the square to be tipped off the drawing when passing back to ink in a neglected line. The edge of the stock immediately under the blade is bevelled off, so that the stock will only touch the board in the centre of the edge, the extreme angle of the board being liable to become indented and untrue.

Moveable Head Square.

The squares already described will only draw lines horizontally or vertically to the edge of the drawing-board, and are used practically only for the *horizontal* lines of the drawing. There are several methods of making the rabbet of the square moveable so as to produce *oblique* parallel lines. One of the most usual is represented in the illustration above, which is the same, in most respects, as the square first described, except that one of the rabbets is formed by a separate piece of wood, which is centred so that it will turn to any angle in relation to the blade; at which angle it may be firmly clamped below by the nut or screw which forms the centre.

This form of square is often found very useful to work parallel oblique lines in any direction. It is also found very convenient if loose drawings have to be laid down on the drawing-board a second time to make additions, or to obtain tracings, it being almost impossible to fasten a drawing down a second time in a true line with the tee-square. In modern practice this square is very little used; it is more general to have a fixed head to the square, and to erect oblique lines with

M

a separate instrument, which will be described in the next chapter.

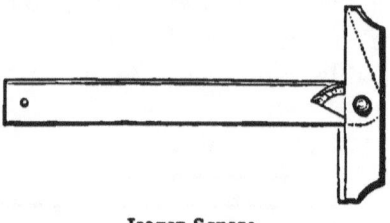

Isogon Square.

The ISOGON SQUARE is the author's plan of making a bevel square. It is constructed somewhat upon the plan of a mechanic's bevel; instead of one side only of the stock moving, the whole of the stock moves, the blade passing up a groove in the centre of it. Upon the blade a protractor is divided, which indicates approximately the angle to which the blade is raised in relation to the horizontal lines of the drawing. The stock has a projecting tongue on the lower part of its edge, which forms a stop, upon which the blade falls when it is required to be used as a tee-square. The advantage of this construction over the last described is that an angle may be set at once by the degrees, and the same angle may be reversed; for instance, if it be set to the angle of one side of a cottage roof, by turning the isogon over, it will produce the opposite corresponding side.

The isogon may be recommended to use as a bevel, but it is not equally serviceable as a tee-square as that described at page 160 in this chapter. It is a very convenient instrument if used supplementary to the other, and no office should be without one.

A very portable form of shifting tee-square, termed

a Manchester square, is represented below. This is found extremely convenient for persons going abroad, as the individual use of three squares may be obtained within less than the space occupied by one. In construction, the stock of this square is an entirely separate loose piece, in the centre of which is fitted a screwed pin, with a very strong clamping nut. Upon

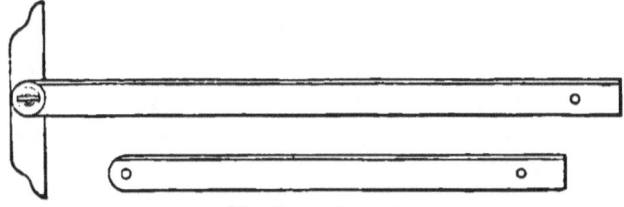

Manchester Square.

the pin are fitted two or more blades, either of which may be screwed on separately to form a square. In using this square the blade is set to the edge of the mounted paper, and firmly clamped. Of course it may not by chance be at right angles to the stock; but this is of no consequence, as a parallelism of the lines is all that is practically required of a tee-square. This square is recommended for portability; it is not so good for general office use as the taper square described page 160, for the reason that the stock by a slight jar may be altered from the angle at which it was first set, and this may occur when a drawing is unfinished, and thereby give considerable trouble.

CHAPTER XXIII.

RULING EDGES FOR PRODUCING PARALLEL LINES—
PARALLEL RULES—ROLLING PARALLELS.

Plain Parallel Rule.

THE ordinary form of a plain Parallel Rule, which is illustrated above, is too well known to need a lengthened description. It consists of two similar straight rules, which have two equal metal bars jointed upon them; the joints forming centres upon which the instrument works.

To ensure accuracy in making this kind of parallel rule, it is only necessary to observe that the distances of the centres in the two bars must be exactly alike, as also the distances at which the centres are fixed upon the rules.

In using the plain parallel rule, one of the rules is pressed down firmly with the fingers, while the other is moved by the centre stud to the distances at which parallel lines are required. Should the bars not extend a sufficient distance for a required parallel line, one rule is held firmly, and the other shifted, alternately, until the distance is reached.

It will be found very awkward to draw a distant parallel line with this rule if it be required vertically over the first line, as the angle of the bars causes the rule to move obliquely; this may, however, be accomplished by throwing the foremost rule over on its centres, so that the bars point outwards to the right instead of to the left, which will cause the rule to move the reverse way, and correct the obliquity of the first movement.

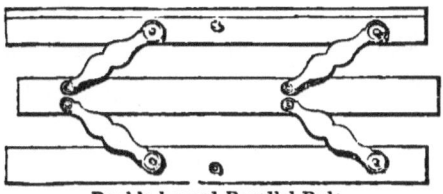

Double-barred Parallel Rule.

The DOUBLE-BARRED PARALLEL RULE, illustrated above, may be considered an improvement on the plain parallel, as the ruling edge moves to a greater distance from the fixed rule, and also moves in a direct line. The principle is the same as the last described, the difference being the addition of an extra rule and pair of bars, which are jointed at the reverse inclination to the first pair. This being more difficult to make than the last described, is seldom as true.

There are two other kinds of what are technically called *bar parallels*—the *cross bar*, and the *double sliding bar*, of which descriptions are unnecessary, as they are almost out of use. There is a considerable difficulty in using any kind of bar parallel, the attempt to hold one of the slippery rules on the paper by mere pressure, while the other is being shifted, requires constant attention, and is at all times an uncertain operation. Another defect of bar parallels is the small advance made by one shifting, which renders the working with them very slow and tedious. For these reasons they are almost out of use, except by lithographers for drawing on stone, and in schools, their place being generally supplied by the set-square and straight-edge, or the rolling parallel, which is more efficient, and, if properly made, a more exact instrument.

Captain Field's Nautical Parallel. This is similar to the plain parallel rule at page 164, but is made in boxwood, and has one of the sides protracted in degrees, and the other in bearings. It is very inferior to that described with protractors further on for the same purposes.

Rolling Parallel Rule.

ROLLING PARALLEL RULES, which are the only parallels used by professional draughtsmen, will produce parallel lines in any direction or at any distance. They move freely, and admit of easy and rapid use. In construction they consist of a solid rule of wood or metal, which is raised a short distance from the drawing upon a pair of wheels of equal diameters. These are united upon a long axle, which revolves in bearings fixed at its ends. Fine grooves are cut round the circumference of the wheels to ensure perfect contact with the surface of the paper. A piece of wood or metal, called a bridge, is fixed over the axle to protect it, which also proves a convenient means for moving the parallel.

Sometimes small ivory wheels with divided edges are fixed by the side of the cut wheels. These will give indication of the distance which the roller has traversed: they are of very little practical use, and as they remove the wheels farther from their bearings they render the action of the rule less solid and exact.

Ebony rolling parallels have, generally, slips of ivory inlaid upon their edges, which are divided similarly to a rule, in tenths and sixteenths of inches, and are occasionally found convenient for drawing details. Drawing scales are frequently put on the edges of parallel rules; they are not very useful. The edge of the rule would be better if left the proper thickness for ruling an ink line, which it cannot be if scales are required.

Small ivory rolling parallels are frequently divided on the edges to form a protractor. These have the roller sometimes placed upon the upper surface, instead of being cut through the rule as in the other kinds. See Rolling Protractors, in Chap. xxix.

The best descriptions of rolling parallel rules are made entirely of brass or electrum, the weight of the metal being of great service in keeping the wheels in constant contact with the paper; they are also more reliable than wooden ones, as with them warping is impossible. The surface of the metal does not touch the paper, therefore they do not soil the drawing. The writer has also made them of vulcanite, which appears to answer well, but is rather brittle.

A little care is required in using the rolling parallel. The left hand should be placed as nearly as possible upon the centre of the bridge, so as to secure nearly equal pressure upon the wheels; and in moving it from line to line the edges should be tipped off the paper, and not scraped along it. If these particulars be observed, very little practice will be sufficient to use the rolling parallel effectively and accurately.

The accuracy of workmanship of a rolling parallel rule is easily discovered by reversing the sides of the rule that are placed upwards, running it down from each edge from a fixed line, and drawing a line at a foot or more distance by each of the edges. If these lines are drawn near to each other, any defect of parallelism in the work, or in the size of the wheels, is detected.

CHAPTER XXIV.

RULING EDGES FOR PRODUCING RADIAL OR VANISH-
ING LINES—THE CENTROLINEAD—ROLLING CENTRO-
LINEAD—EXCENTROLINEAD.

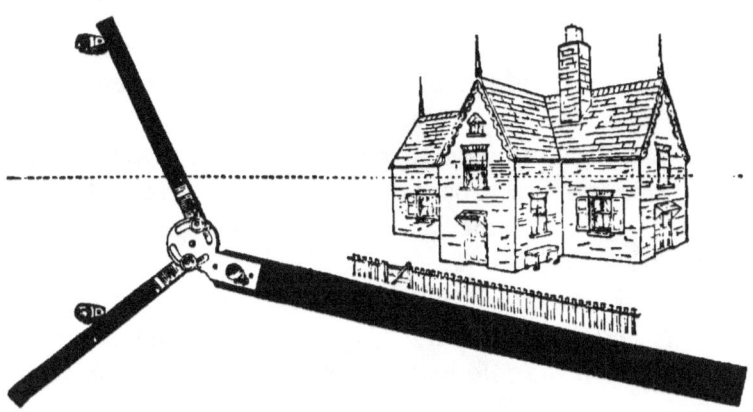

Nicholson's Centrolinead.

THE CENTROLINEAD, which is illustrated above, was invented by Peter Nicholson, a man of great geometrical
ingenuity. It is used entirely for drawing lines from
an imaginary distant centre, called, in perspective
drawing a vanishing point. This point is frequently
at such a distance, that it would require a very long
drawing-board and straight-edge to produce the
vanishing lines, were it not for this very convenient
instrument, which will work radial lines from any distance without requiring greater surface to work upon
than that occupied by the finished drawing.

In construction the centrolinead consists of a long

plain rule, upon which are jointed two arms in such a
manner that they may be set to any required angle.
The two arms are first jointed together in the manner
of a mechanic's twofold rule, having a pin or stud projecting from the centre of the joint. This stud fits
into the centre of the plate, which is carried over from
one end of the long rule, thus rendering the rule and
the two arms adjustable upon one common axis. In
the plate are two grooves or slots, concentric with the
axis of the instrument. Through these grooves two
milled-head screws pass to the arms beneath, forming
means of clamping them at any required angle to each
other, and to the rule. The plate described is connected upon the rule by a milled-head screw and steady
pins.

It will be observed, on examining the illustration,
that one edge of the long rule comes in a line with the
axis of the centrolinead. In this position the instrument, as now fixed, will be adapted to produce vanishing lines from the *left-hand* side of the drawing only,
as it is necessary to have the edge of the rule which
leads to the axis towards the top of the drawing. If it
were now required to be changed to produce vanishing
lines from the *right-hand* side of the drawing, it would
be necessary to take out all the milled-head screws, and
to turn the circular plate the reverse side upwards. In
this position, when re-fixed, the upper edge of the rule
would lead to the axis upon the right-hand side of the
drawing.

In drawing with the centrolinead, the arms are
pressed continually against two *studs*, which are fixed
at a distance apart upon the edge of the drawing
surface. The stud is a piece of metal with a rounded

edge, rising about a quarter of an inch from the surface of the drawing-board. It is fixed to the board by
a pin projecting from its under side, and a second pin
passed through a thin flange upon the side farthest
from the drawing.

One method of setting the centrolinead for use, which
is the manner recommended, is as follows. After
drawing the horizontal line for the intended perspective
drawing, which is generally done by the tee-square, a
vertical line has to be drawn at right angles to it, up
the side of the drawing-board from which the vanishing lines are wished to be produced. Upon this line,
at equal distances, generally about eight inches from
each side of the horizontal line, are to be placed the
two studs which are intended for the arms of the centrolinead to slide against. These studs are fixed in
position by pressing down the pin which projects from
the under side in the point of distance set off on the
line, and afterwards firmly secured by a drawing pin
through the flange. The upper or axial edge of the
rule of the centrolinead is then placed along the horizontal line, and the arms, the screws of which have
been previously loosened, are each brought to one of
the studs, allowing the arms to take about the angle to
each other thought to be required to produce the desired distance of vanishing point; in this position the
arms are to be clamped. It is then necessary to try if
the centrolinead will correspond with the line which
forms the top of the building, or other object intended
to be placed in perspective, which is either sketched
by judgment or drawn according to the rules of perspective. This is done by moving the rule up from the
horizontal line, always keeping the arms in contact

with and sliding against the studs. Should the vanishing point that would be given by the centrolinead as now set appear too near, it will be necessary to put the rule back on the horizontal line, from which it has always to be set, unclamp the screws of the arms and press the rule back against the studs, keeping it still on the horizontal line, so as to flatten the angle of the arms the amount thought to be required; then clamp the arms again, and make another trial.

If the vanishing point appear too far, the arms will require setting at a more acute angle. It is best in all instances to make a mark at the side of the end of the centrolinead to show its position before alteration, to ensure having about the difference from the last setting thought to be required. When the instrument is once set, it is right for all the vanishing lines from one point, and it is advisable not to shift it until the drawing is quite finished.

The above description appears much more difficult than it will be found in practice, as after using the centrolinead for a few perspective drawings, the angle at which the arms should be set for any particular drawing becomes so familiar, that it may be judged sufficiently near for the first trial, or a slight alteration of this, to suffice.

There is another method of setting the centrolinead, much more simple, but, practically, not so good as that above described, as it only by chance allows the full distance of action and firm bearing of the arms. It is also difficult by this method to keep the studs off the drawing. The plan is, however, in common use, and may by considerable judgment in guessing the most convenient angle, succeed very well; it is as follows:

The arms of the centrolinead are fixed by the clamping screws at any angle, according to judgment, being particular that they are at about equal angles with the rule. The rule is placed along the horizontal line with the arms passing a little over the edge of the paper; a pencil mark is now made along the outer edges of the arms. The rule is then moved and placed on the line sketched for the top of the building, or other object of which the perspective drawing is intended, and the arms are put over the lines made from the first setting; so that by drawing lines outside the arms again in this position of the instrument, the first lines will be intersected it two places; in these two intersections, the studs are to be placed, and the whole is ready for work.

There are two other methods of setting the centrolinead, by calculation of angles and by reversing intersections. They are more tedious than those described, and the introduction of these methods would only tend to confuse.

To obviate the difficulty of setting the centrolinead when the intended distance of vanishing point is known, the author has sometimes divided the edge of the plate with lines indicating distance of vanishing point in feet, to be read off by an index-line on each of the arms, the studs requiring to be uniformly at eight inches distance from each side of the horizontal line. This division will be found convenient, and its extra expense very trifling.

In any office where many perspectives are drawn, the draughtsman will find it very expeditious to have a pair of centrolineads, one made to work towards the right hand and one to the left; as the centrolinead once fixed to a vanishing point should not be shifted until the drawing is completed.   A pair of centrolineads, each made to one hand only, will be less expensive than two centrolineads made to work to either hand.

In default of two centrolineads, it is best to have the largest drawing-board at command, and to place the intended drawing near one end, as this will admit the use of a straight-edge for one vanishing point.   The great convenience of the centrolinead is that it requires no space to work upon except that of the paper upon which the drawing is to be made.

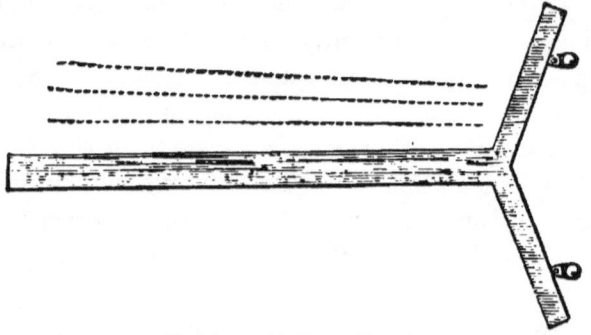

Shuttleworth's Centrolinead.

From the second method described of setting the centrolinead, it would appear that the instrument fixed at an arbitrary angle is made to answer for any distance of vanishing point, so that a centrolinead might be constructed with a fixed angle to move against the studs, instead of an adjustable angle, which being more simple, would be much less expensive.   This kind of centrolinead is in use; it is known as Shuttleworth's centrolinead.   It is generally made in the form of a tee-square with a long stock, the back of which is cut to an angle similar to the angle of the arms of the ordinary centrolinead when set.   It is, however, a very imperfect instrument, its defects being, that if a moderately near vanishing point is required, the intersection from marking the angle brings the studs nearly close together, leaving no steadiness in the rule ; and if the vanishing point is distant, the intersections, and consequently the studs, are so far apart that the rule has no range of movement. It is described in these pages, being an instrument in use, that its defects may be pointed out, and that no draughtsman who may be purchasing one, should anticipate deriving from it the perfect uses of the centrolinead, as it is really perfect for one distance or vanishing point only, although it may be tortured into use for others.

Shuttleworth's centrolinead is occasionally made to work on curved edges instead of studs; this is little improvement, except that it is useful for very small perspectives, such as may be required for engravings or lithography, in which instance the centrolinead, being very small, may be cut out of a piece of vulcanite, the arms at their outer extremity being made to terminate with projecting round edges. About half a dozen curves to form the templets will be sufficient variety to give distances of vanishing point available for almost any required perspective. Each templet, for wood blocks especially, should have a fillet turned down, so that it may be clamped or wedged to the edge of the block, to hold it in the required position.

In perspective drawing, the studs which are fixed at the side of the drawing for use with the ordinary centrolinead, will be found awkwardly in the way of using a tee-square from the side of the board in the usual manner. This may be obviated by using a short tee-square with parallel blade from the bottom of the drawing-board only. When horizontal lines are required, they may be produced by applying the set square.

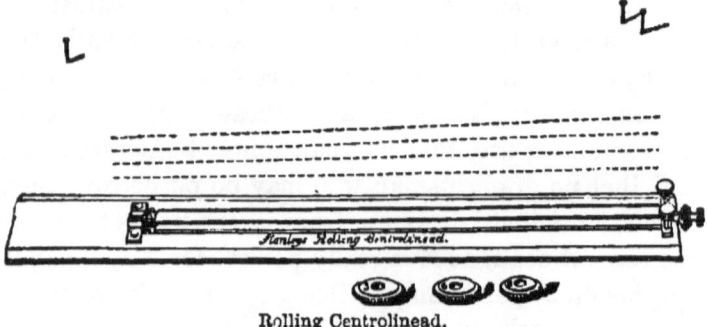

Rolling Centrolinead.

There is another kind of Centrolinead, called a Rolling Centrolinead, which, although in all respects not so perfect as the one just described, has the merit of being much more portable, and may be used from either side of the drawing without changing the parts. It may also be used as a parallel rule, and is sufficiently exact and more convenient for many purposes. It has gone almost out of use, perhaps on account of some little improvement the original instrument required in the details of its construction. As the author has constructed it, it resembles in most respects the rolling parallel rules described in the last chapter, except that one of the wheels, which is brought to the end of the rule, is made to take off, to be replaced by a larger one : the bearing, also, next the changeable wheel is made to rise up by adjusting screws, so that the rule may retain its horizontal position, although the wheels are of different sizes.

There are ten loose wheels supplied in the case with the instrument,—one of equal diameter with the fixed

wheel, to form a parallel, and the others to attach to
the instrument to produce radial lines respectively
at 3, 4, 5, 6, 7, 8, 9, 10, and 12 feet distance from the imaginary vanishing point.

*In using the instrument, after the required wheel is
attached and the bearing adjusted,* it will be necessary
to observe what has been said upon using the rolling
parallel, with a further precaution, that the rolling centrolinead must be kept constantly in one curve of position upon the drawing. This will be evident from the
fact that the wheels are made to give a distance of
vanishing point from the centre of the ruling edge ;
thus, if the rule be shifted longitudinally, the relative
distance of vanishing point will be changed. The most
convenient way of keeping the instrument in position
is to have a constant reference line. The manner recommended would be to roll the centrolinead from its
intended position on the horizontal line up to the top
of the sheet of drawing-paper ; in this position to draw
a line along the top edge and one end of the instrument, and to drive in an ordinary household pin near
each end of the lines, also to place a pin in the end
line. By afterwards placing the instrument in the
angle formed by the three pins, its position, in reference to all the vanishing lines of a perspective drawing,
may be instantly restored. The positions of the pins
are shown in the illustration.

This centrolinead is in no way so convenient to use as
Nicholson's. It has one small merit only, portability.

Centrolineads offer convenient means of working to
the rules of linear perspective, and are in every way
equivalent to other means of producing vanishing lines.
Drawings produced by such aids are useful for giving

N

an idea of an intended work, building or other. But the best results of linear perspective can scarcely be held to be artistic. This occurs, not from defect of the instrument, but the principle. In making a linear perspective of a building the eye is supposed to be placed in exact opposition to the most prominent angle, and this angle from the width of the face of a building, compared with its depth, has very frequently to be placed near one side of the drawing. Now, in looking at any drawing, we take our view from the centre of it, and by the principles of linear perspective applied in other instances, all horizontal lines should vanish from this central position. It is also the fact that all vanishing lines should tend to vanish in some degree towards both hands from this point. Therefore a line, although vanishing by perspective rules all to one hand, if it passes the centre of the picture, should vanish to the other hand also, in a certain degree; therefore, if the line were above the point of sight and pass the centre, it should *round upwards*. And if we draw the line straight, as it is done universally by the rules of perspective, such a line appears to *be hollow*, and the perspective angle, now assumed to be towards the side of the drawing, appears much too sharp for natural effect. This principle of drawing gives professional architectural drawing a most unnaturally stiff effect, as may be observed by looking over those exhibited annually at the Royal Academy, that is, to any one not used to the system. For the drawing to look real, all the upper lines, and indeed all lines that vanish past the centre of the drawing, except, of course, the horizontal line on the point of sight, should be curved, but the upper *ones* especially so.

Some professionals to whom I have mentioned this matter do not appear to see it as I do. It appears to me to be easily explained. Thus, that if we stand facing a long high wall, the part that is immediately opposite to us appears level; but as we look away on either side - the perspective vanishes, and should be represented to

do so. The wall would appear to us, in this case, if our view could include the whole of it, as a very flat hyperbola. It is also clear that if we observe from any point of a vanishing line that passes the centre of our view, that this must appear to do so also. For if the vanishing perspective line were extended until it reached behind us, the same conditions would occur as though we viewed the long straight wall directly and obliquely. It might be thought upon this principle that heights and vertical lines should also vanish into the distance. And by equivalent laws to those applied to linear perspective on the horizontal, this would certainly be the case. But *no correction* is needed for the vertical lines. The form of the retina and crystalline lens of the eye possibly corrects the difference of vertical angles in some mysterious way of which the writer has never met with satisfactory explanation; so that if we look at a straight diminishing column, it appears to the eye hollow or smallest in the centre, which is presumed to be from over perspective correction in the eye. Therefore, columns are made rounding outwards, to produce the effect of equal diminution. But if we look at a wide parallelogram or building, all the apparent incurvation disappears, so that we must infer that no correction is needed in the perspective for this, although it is needed in the horizontal, where the eye takes correct cognizance of the angle subtended.

I should not introduce this matter, only that it may be remedied very much by simple means, and linear perspective drawings even be made to look artistic. To do this, in the first place it may be observed, that the eye seldom regards more than the outline for general perspective effect, and the most important part of this as it comes strongest in view is the sky line. Now it is only slight curvature that is required in the lines that pass the centre, and the lines drawn straight with the pencil may be conveniently inked in with a rule that has a slight curvature. The curvature should be such that the rule should have a gradually increased curvature from one end to the other, the curvature vanishing off to a straight line for the distance—in fact, be hyperbolical. By shifting the rule to different parts of its curvature, the principal horizontal lines, much above or below the point of sight, may be drawn, and the remainder of the perspective be finished with the centrolinead without apparent defect. Even lowering the angle, if the vanishing lines are carried across the drawing to the opposite side, adds much to the true perspective effect. In sketching from nature these faults do not, of course, occur. The best argument for making all the lines straight is, that the perspective may be considered as a model, where the eye takes the whole in at a glance, or, that the object may be assumed to be in extreme distance. If this argument holds, it is right; but the

defect of straight line linear perspective of buildings
generally appears to be, that they look too much like
models, when by means suggested they might be made
to look like buildings, or natural objects delineated by
an artistic hand.

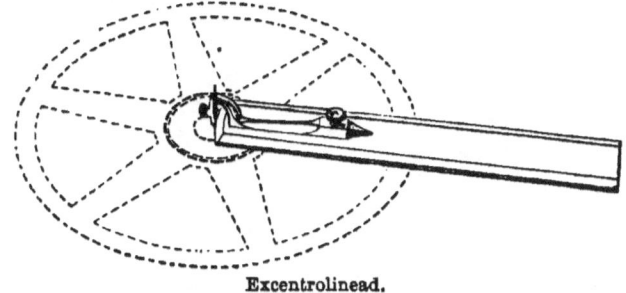

Excentrolinead.

The EXCENTROLINEAD, illustrated above, is used
mostly on the Continent for drawing excentro-radial
lines, such as the sides of the taper arms of cog-wheels,
circular-saw teeth, etc.; it is a very inexpensive instrument, and will be found useful for some descriptions of mechanical drawing. In detail it consists of a
small rule of ivory about six inches long, with a moveable arm jointed upon it; the arm carries a needle at
its point, which can be set to any angle in relation to
the line of the edge of the rule.

*To use the excentrolinead for drawing the arms of
a wheel,* the general position of which is shown in the
illustration: clamp the arm of the instrument at the
angle required in relation to the centre of the wheel;
place the needle point at the end of the arm in the
centre from which the wheel was struck, and the edge
of the rule will give one side of all the arms; for the
other, the point must be reset by the scale at the end,
which takes only a minute to do.

CHAPTER XXV.

RULING EDGES USED TO RAISE ANGLES FROM THE EDGE
OF ANOTHER INSTRUMENT—SET SQUARES—SLOPES
AND BATTERS—LETTERING SET SQUARES—SECTION-
ING SET SQUARES—ISOGRAPH, ETC.

Set Squares.

THE SET SQUARE is perhaps the instrument of all
others most constantly in the hands of the draughtsman. It is employed to erect all perpendiculars, and
several other frequently required angles and parallels,
by making use of the edge of the straight-edge, tee-
square, or parallel rule, as a base.

The most simple form of set square consists of a
triangular piece of thin wood, one of its angles being
uniformly of 90 degrees, or right-angled. The com-
plementary angles are varied to suit the purposes to
which they are employed, the most general being the
45 and 60 degrees.

The material very frequently employed for set
squares is pear-wood, which from its uniform density
produces a smooth edge in every direction of the grain;
it has, however, a fault inherent in all wide pieces of

thin wood, which is, that it constantly warps or shrinks, leaving its surface and edges untrue.

The author, after several experiments in seeking some suitable material for plain set squares, discovered that Goodyear's vulcanite, which is a patented preparation of india-rubber, of which the best qualities are manufactured in North America, possesses all the qualities desirable for set squares to be used in temperate climates. This material is considerably harder and tougher than any kind of wood; it is impervious to moisture, consequently it may be kept clean, if required, by washing, and it will not warp or get out of truth under any ordinary circumstances. Metal set squares possess some of these qualities, but they are heavy, and soil the drawing by condensing the perspiration of the hand.

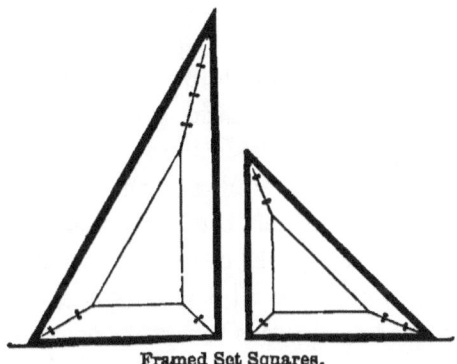

Framed Set Squares.

FRAMED SET SQUARES, as illustrated above, are the best kinds of set squares for large or moderately large sizes, or for use in hot climates. They are generally made of mahogany, and edged with ebony. Although made of wood, it is in such narrow strips that the contraction of the fibre is inconsiderable. These set squares also retain their angle very correctly, from the fact of the grain of the wood running longitudinally on all sides. Being open in the centre is also an advantage, making the set square lighter, and obscuring less of the surface of the drawing than the solid form. They require very careful workmanship; the angles should fit perfectly, and be united with a tongue of metal which should be riveted through to the surface; made thus, the angles will be as strong as the sides.

In using the set square to produce perpendicular lines, it must be held with constant light pressure upon the edge of the tee-square or parallel rule, whichever is used; the middle finger of the left hand being placed in the hole if it is a plain set square, or upon the inside edge of the bottom side if a framed one.

There is a difference of opinion amongst draughtsmen as to which side of the set square the perpendicular should be drawn; some holding that the vertical edge should stand to the right hand, and the pen be drawn downwards towards the body, and some that the edge should stand to the left hand, and the pen be drawn upwards. Upon this subject the author can only offer an opinion, as there is a great deal of experience in favour of either way; and after all it is only a matter of practice. It, however, appears to him in some respects inconvenient to place the vertical edge of the set square to the left hand, particularly from the hands having to be crossed, and the right hand being placed over the left, which appears awkward, and tends to obscure the light from the line; it also causes the body to reach farther over the drawing than is necessary; whereas, placing the vertical edge to the right hand and drawing the pen downwards, appears more elegant, although, perhaps, it requires a little more practice to acquire an easy and exact facility of using the instrument in this position. Many draughtsmen work to either hand.

Gallows Square.

Ship draughtsmen very commonly use long straight-edges, which are clamped to the bottom of a drawing, and erect a perpendicular, with a long form of set square, from twelve to thirty inches, termed techni-cally a *gallows square*. For short horizontal lines an ordinary set square is used on the edge of the gallows square.

Besides the use of the set square for producing right angles, the more acute angles come very frequently into requisition. The 45 degrees is useful for uniting angles, or preparing plans for perspective drawings. The 60 degrees and 30 degrees are used to represent the hori-zontal lines of isometrical perspective, sides of the hex-agon, etc.

Set squares may also be conveniently used as a parallel rule, by sliding them either upon each other or along the edge of a straight-edge or tee-square. By sliding the set square along the edge of a scale, parallel lines at equal distances may be drawn, which is the most convenient method of drawing sectional lines on mechanical drawings accurately. In every respect the set square is much more convenient for drawing short parallels than the nearly useless bar parallel rule, already described, as generally supplied with cases of drawing instruments.

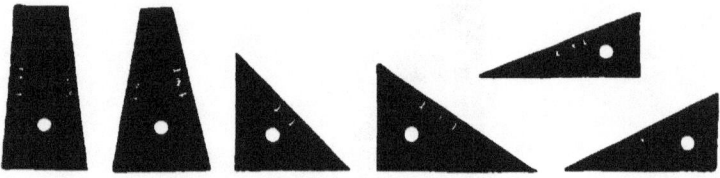

Slopes and Batters. ·

SLOPES AND BATTERS are a class of set squares arranged by the writer to form a set of angles constantly recurring in railway engineering. They consist of six or eight pieces, and contain the ordinary slopes for sections of earthworks of 1 to 1, 1½ to 1, 2 to 1, 2½ to 1, 3 to 1, etc.; also the batters for walls and rocks of 1 in 4, 1 in 5, 1 in 6, 1 in 8, 1 in 10, 1 in 12, etc. They have been much approved of by civil engineers, and are very inexpensive.

Roof Pitches.

Slopes similar to the above, technically termed *roof-pitches*, are used by architects. Six of these form a set, viz., one-seventh, one-sixth, one-fifth, one-fourth, one-third, and one-half; these give the established slopes for roofs, and are very convenient.

Nut Angles.

Two slopes suggested by the writer are very convenient for mechanical engineers; these are illustrated above. They give the angles of hexagon nuts two ways, and allow the nut with care to be inked in without the slope passing over a wet ink line; to do this the point of the slope has to be inserted into the top angle of the nut; it is also better to do the left hand top side line' first; the cost of the two in vulcanite is trifling.

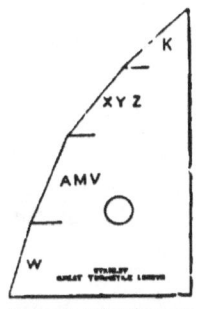

Lettering Set Square.

The LETTERING SET SQUARE is a form of set square which the writer invented to give the oblique angles which occur in the alphabet. It may sometimes be used by draughtsmen who have a difficulty in lettering titles of plans, etc., if the very useful substitute, stencil plates, be thought too expensive. This set square is

shown in the engraving; it is used upon the tee-square
or parallel rule, which is adjusted to bring the portion
required between the lines forming the top and bottom
of the letters.

Sectioning Set Square.

A very convenient apparatus connected with the set
square the writer has seen used on the Continent to
produce section lines on mechanical drawings. It consists of a kind of lever, which is raised by a spring and
stopped by an adjustable screw, so that it will only
move a certain distance by pressing the finger upon it.
A kind of foot with a chisel edge, constructed so that
it will bite upon the surface of the drawing, is jointed
to the lever in such a position that the pressure of the
lever will cause the edge of the foot to bite the paper,
and form a fulcrum by which the set square advances
consecutively the required distance, according to the
adjustment of the screw.

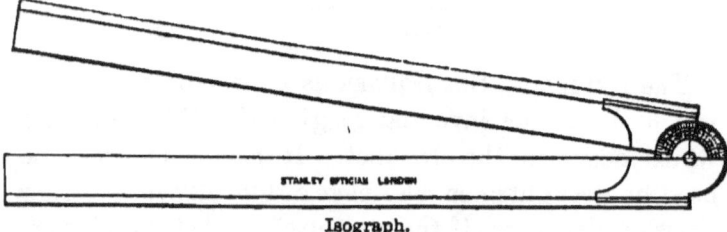

Isograph.

In architectural and mechanical drawing, corresponding angles are frequently required, which are not of sufficient importance to be worth having a set angle for their production. For these purposes the Isograph will be found very convenient. In construction it resembles a mechanic's two-fold rule, the joint being made stiff and of extra size; the middle plate of the joint does not come through to the outer edges, therefore the edges may be bevelled down to a proper thickness for producing an ink line. The head of the joint has a protractor divided upon it, so that the two edges of the rule may be set to any angle approximately.

The isograph is intended to be used in a similar manner to a set square, as an accessory to the teesquare, principally for such purposes as slopes of roofs, spires, cones, and other instances where a corresponding angle is required to right and left hand.

CHAPTER XXVI

RULING EDGES FOR PRODUCING CURVED LINES—RAIL-
WAY OR RADII CURVES—SHIP CURVES—WEIGHTS
AND SPLINES—CURVE BOW.

For the production of curved lines, which are constantly recurring in every description of geometrical
drawing, it has been found in practice, that nothing is
so convenient as the edge of a thin piece of shaped
wood, which acts as a templet to form the required
curve. These templets, which are of several different
kinds, are called curves. They are useful for many
purposes in drawing, both mechanical and ornamental.

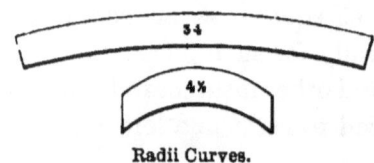

Radii Curves.

Railway or Radii Curves are thin pieces of wood
or cardboard cut with a beam-compass into arcs of
circles of radii from 2 to 250 inches; they are generally
made about 2 inches wide, and from 3 to 18 inches
long, the length increasing with the radius. They are
packed in boxes of 25, 50, or 100 curves.

Radii curves are principally used for railway plans,
for which purposes they are found most convenient if
the outer and inner arcs are made of the same radius.
If made of wood the curves are seldom correct, from
the tendency of the grain to draw the knife, however

rigid the cutting beam; also, from the wood springing to greater or less radius immediately it is cut. For these reasons cardboard or metal curves are found much more correct. Cardboard forms naturally a very soft ruling-edge, which is particularly disadvantageous for producing a clear line; but the edges may be very much improved by indurating the surface before it is cut with a solution of shellac in alcohol, which will render it impervious to moisture and dirt, and produce a moderately firm edge. Thus prepared, cardboard forms a perfect and economical curve, sufficiently durable for all practical purposes, with the especial merit of standing true in all climates. Vulcanite, mentioned for set squares, makes also very excellent curves.

Curve Radiator.

For erecting perpendiculars or radii from railway curves, the small instrument illustrated above answers perfectly. It is cut out of a piece of vulcanite, and goes into the box with the curves. It will work from either side of the curve, by placing the hollow end against it.

SHIP CURVES, of which illustrated specimens of the forms follow, are used for drawing in the framework, casing, or general curved lines of ships or boats. They are made of thin pear-wood or vulcanite, the

forms being those which occur most frequently, and
are nearly all mathematical curves derived from the
helix, ellipse, or parabola. A single curve will seldom
give the entire curvature of a rib of a vessel, or other

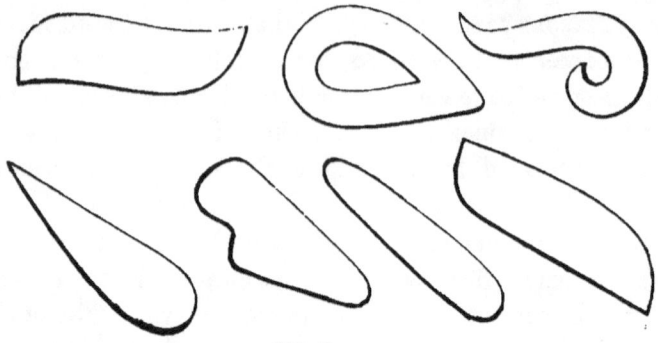

Ship Curves.

complete line, but it may give a part of the curve, a
second or a third being used to complete it. In this
manner nearly all the curved lines which occur may be
produced with moderate accuracy. The set of curves
of this description, as used in the Admiralty depart-
ments and the principal shipbuilding firms, are forty in
number; there is a supplemental set of forty, making
eighty in all, which are used by many large firms.

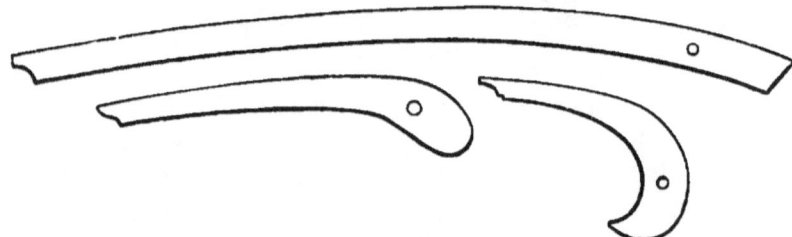

French Form, Ship Curves.

The French use somewhat different forms to the above
described, the curves being made to use internally as

well as externally; they are less comprehensive than the English forms, but are of very excellent lines. The set consists of fifteen curves, three of which are illustrated. Although the French form of ship curves is used principally by the ship architects, they are also the most useful class of curves for the mechanical engineer.

FRENCH or IRREGULAR CURVES are used for rounding angles, ornamental parts, and various curved lines upon geometrical drawings. They embrace a great variety of patterns, a general idea of which may be formed from the illustrations given below. These curves, by piecing out, may be made to form any curved figure of moderately large dimensions.

French Curves.

When similar forms are to be placed to the right and to the left hand of the drawing, it is customary to mark the edge of the curve, as far as required to be drawn from, on one hand, and to turn it over to trace the same portion to the reverse hand, the marks preventing error in quantity of form. Draughtsmen will very frequently make one or two pet curves answer all purposes, by using a very short piece at a time. It will be generally found much more expeditious to have a set of about twelve of the useful patterns, the whole costing about five shillings. These will be found to

O

answer all the requirements of the mechanical draughts-man. Curves are sometimes made in metal, but they are better and cheaper in wood or vulcanite.

In the ordinary form of French curve the curved lines are too flat for architectural drawing, which is

Architectural Curves.

mostly made to an eighth or quarter-inch scale, the curved lines being required for trusses, caps, mouldings, arches, etc.   The author arranged four curves for these purposes, the forms of which, as illustrated above, will be sufficient description to suggest their various uses.

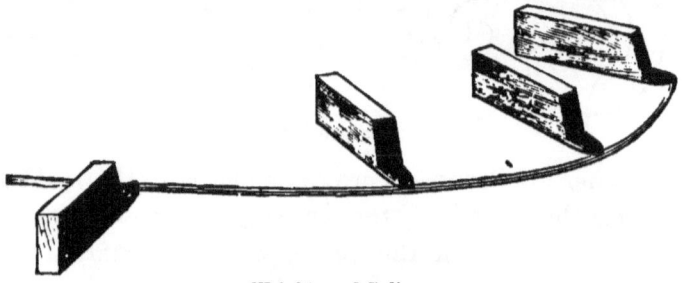

Weights and Splines.

WEIGHTS AND SPLINES, or *Penning Battens*, as they are sometimes called, are used to form figures of irregular large curvature, which they delineate more gracefully than the curves already described : they are used principally by ship architects.   The set of weights and splines generally consists of six weights, and twelve to fifty splines.   The weights are usually made of lead, and neatly covered with mahogany; they have one end brought out in the form of a wedge, the thin edge standing up vertically upon the drawing. The splines, or penning battens, are thin pieces of lancewood or red pine, from eighteen inches to eight feet long, and from three-eighths to one-eighth of an inch square, parallel throughout; others are made to a diminishing taper from end to end, others diminishing from the centre to the ends, or from the ends to the centre in varying proportions. They require making with great care by a practised workman, or they will not give a fair curve. The wood will occasionally spring during working, but if it falls to a fair curve this is of little consequence, whereas any inequality in substance will make a *cripple*.

The *method of using the weights and splines* is to place the points of as many weights as are thought to be necessary in the position of the required curve. The spline is then sprung round the points of the weights, or the points of the weights may be placed upon the spline, as most convenient to hold it firmly. The drawing-pen is drawn round the spline to produce the curve, with as light pressure against it as possible. The cranked or curve pen answers well for this.

Curve Bow.

Several instruments have been made upon the principle of the weights and splines—the weights, in most instances, being dispensed with, and the spline fastened in various ways. One of these schemes that may be useful to the engineer is illustrated above. This is an instrument intended to give the form of the under side of a metal beam.   It consists of a moderately wide piece of wood, fitted with two thin brass links to slide upon it.   In the centre of the bar is a milled-head screw, which passes through the flat way of the bar. A thin spline of wood is brought through the two links and under the screw.

In using the instrument, the two links are put in the position on the bar to represent bearings of the beam, and the screw, which represents weight, is brought down on the spline to the distance of the required depth of beam.   The curvature thus produced gives the form of a beam to support the greatest weight in the centre, or other position upon which the screw may be brought down.   The instrument is also useful for similar curves for other purposes; splines of varying width, thickness, and length may be adapted to the same bow, and give variety to the curvature.

WILLIS'S ODONTOGRAPH.—This is an instrument which is made either of brass or cardboard, for striking the curves of the teeth of wheels, by compasses, from a kind of divided bevel, which is set at a fixed angle of 15 degrees to the radius.   It has also engraved upon it, if brass,—or printed, if cardboard,—scales for the proportions of tooth to space, and of the position of the pitch line for teeth of from a quarter to three and a half inches pitch.   A full printed description, with tables for setting, is given with the instrument when it is purchased.   Description would occupy much space; it is only mentioned here to call attention to the existence of such an instrument, used in practice by a few draughtsmen.

CHAPTER XXVII.

GENERAL DESCRIPTION OF DRAWING AND MATHEMATICAL SCALES—MATERIAL, DIVISION, ETC.

DRAWING SCALES are rules for measuring or setting off quantities upon geometrical drawings. The drawings are generally made of less dimensions than the actual objects which they represent, the scale of the drawing being a diminished rule in the like proportion.

Mathematical scales are scales of proportions deduced from calculations, or the qualities of the dimensions of an object of definite form, as a circle or cube.

Scales are generally made of boxwood or ivory. The boxwood most suitable for the purpose is rather small, live Turkey wood. It should be of a clear yellow colour, and of dense waxy grain. Soft, inferior wood soon becomes dirty in use, and the divisions upon it appear woolly.

The ivory suitable for scales is of two distinct kinds —the white opaque ivory, principally imported from the eastern coast of Africa and the Cape, and the transparent, called green ivory, from the western coast of Africa. White ivory is preferred by many draughtsmen. It is the least expensive, shrinks less, and has the great advantage of showing divisions and figures much more clearly than the green ivory. It has one defect; it turns yellow after a few years' exposure. Green ivory is very transparent, of a dull, heavy colour; it does not show the divisions very distinctly until it has been some years in wear, when it becomes of a pearly whiteness, which is unchangeable.

Besides these distinct varieties, ivory is occasionally imported possessing many of the qualities of both varieties. There is a semi-transparent white ivory, of which we receive a very small quantity from Ceylon, of very fine quality for mathematical purposes. Also from Angola we receive a very fine white transparent ivory. The quality of all ivory varies in different parts of the tusk, technically *tooth*, the centre being generally the best.

The great impediment to the more universal employment of ivory for scales is that it constantly shrinks. This principally occurs immediately after the ivory is sawn from the *tooth*, but it does not cease entirely for many years, if at all. A twelve-inch scale taken from near the centre of a green tooth will shrink the thirtieth of an inch in five years, above half this quantity occurring in the first three months. White ivory shrinks in a much less proportion, from its containing a greater amount of mineral matter.

The shrinkage of green ivory may be prevented, in a great measure, by boiling the ivory scales for about twenty minutes when they are first sawn out. They will then require six months' seasoning in the air, after which the shrinkage will be almost imperceptible. This process bleaches the scale and renders it rather opaque, but the ivory does not afterwards lose its whiteness.

The only materials employed for drawing scales, except boxwood and ivory, are metal and paper. Metal scales are expensive, and soil the drawings.

Paper scales are supposed to possess the advantage

of shrinking and expanding with the drawing. This, however, is not the case when the paper is stretched by being glued to the drawing-board, which it should be for any important drawing. Paper scales present a soft, ragged edge, difficult to work from with any degree of accuracy; and they soon become dirty and obliterated, and are altogether the most expensive, if wear is considered.

Vulcanite has been used a little for scales; it is affected very much in length by changes of temperature, therefore it is unfit for them. It is preferred by a few, who state that it is soft to look on by night, and less trying to the eyes.

It is very important that the divisions upon the drawing scales should be accurate, as the work that has to be executed from the drawing is very frequently fifty to one hundred times larger than the drawing, therefore errors in dimension may be increased in the actual work fifty or a hundred-fold the error in the scale.

The ordinary method of dividing scales by copying off the divisions with a dividing knife and square from a pattern, is in no way so accurate as is absolutely required, although a skilled workman attains, considering the method employed, great proficiency and surprising rapidity of execution. Still the mechanical sameness of the work is too tedious to ensure the great attention constantly required; and, after all, he relies upon a pattern frequently divided in a similar manner, and inaccurate.

From the above considerations it is important to the draughtsman to be acquainted with some method of testing his scales. An inaccurate scale will often give inconceivable trouble before the fault is traced to its true cause. The following suggestions are simple means.

Set a pair of spring dividers to measure off a few divisions, and try if the same number of divisions correspond in quantity on different parts of the scale, particularly near the first ends.

For examining a closely divided scale, the accuracy of the work may be judged of pretty correctly by comparing the consecutive divisions by the naked eye. Of course, if badly divided, the spaces will appear unequal.

The best method of testing a scale, if two scales are at hand whose divisions are some multiple of a like quantity, is to lay the scales, edge to edge, on a flat surface, and to observe whether the lines look continuous where the scales read into each other, which on most scales will occur at every inch. The total lengths of the two may also be compared with advantage. Scales are most frequently inaccurate near the ends, where it is most important that they should be accurate.

It has always been found a great difficulty to divide boxwood and ivory by mechanical means. The author considered the accomplishment of this object so important that he devoted several years to the construction of a machine that should perform every description of straight line dividing, applicable to the standards of all nations, bearing in mind the important consideration of producing the work at commercial prices. This he was able to accomplish only by many experiments, and after frequent failures in mechanical detail. A description of this machine would occupy too great a space in a work of this class, devoted more especially to the actual drawing instruments than the tools by which they are produced.

In modern practice, the scales mostly used are all divided to the edge, which is made thin, to enable the quantities to be marked off with a fine pencil or a pricker. Dividers and compasses are used occasionally upon scales to take off small quantities, as the thickness of material, or the radius of a circle; their frequent use wears away the divisions of the scale, and should be resorted to as little as possible.

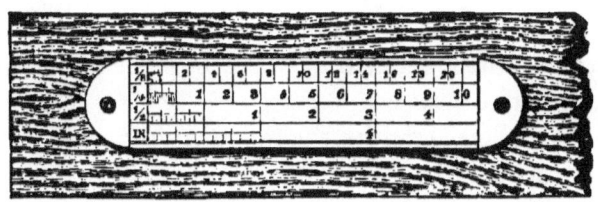

To obviate this, the writer has made small ivory tablets for the scales mostly in use. These are very inexpensive. They may be attached to the tee-square or parallel rule by two screws. The material being ivory they read much clearer than the fine divisions of the boxwood scale, which in mechanical engineering establishments become very dirty and dim. They also save the wear on the scale. The same form of tablet may be fixed to the eighteen-inch boxwood scale. The one represented in the engraving above is for $\frac{1}{8}$, $\frac{1}{4}$, $\frac{1}{2}$, and 1-inch scales. For mechanical engineers $\frac{1}{4}$, $\frac{1}{2}$, 1, and $1\frac{1}{2}$-inch would be more useful.

The three illustrations given on next page represent the cross section of the thin-edged drawing scales in general use, termed respectively flat, oval, and triangular scales. The flat section is to be recommended, as it lays most perfectly to the surface of the drawing. The oval section has the merit of being easily picked up off the drawing, by tipping it in placing the finger on either edge. The triangular section contains six fully divided scales, which are not so easily read off

as the flat section. They are not much in use, nor much recommended, as the cost of one is equal to that of three flat scales.

The ordinary lengths for all drawing scales are a little over six or twelve inches. These are practically too short for scales to be applied to the large drawings which are made in modern practice, mostly upon double elephant or antiquarian paper, as all divisions should be taken with one measurement, if possible. For large drawings generally, an eighteen-inch scale will be found the more convenient and exact, and for large detail drawings a twenty-four-inch scale is not too large. Many engineering firms use thirty-inch for details, to ensure single measurements. For metal-work drawings occasionally contraction scales are used, which are made the excess of the contraction of iron above true dimensions.

CHAPTER XXVIII.

DESCRIPTION OF SCALES IN COMMON USE—CHAIN
SCALES AND OFFSETS—ENGINEERS AND ARCHITECTS'
SCALES—MARQUOIS' SCALES—MILITARY SCALES.

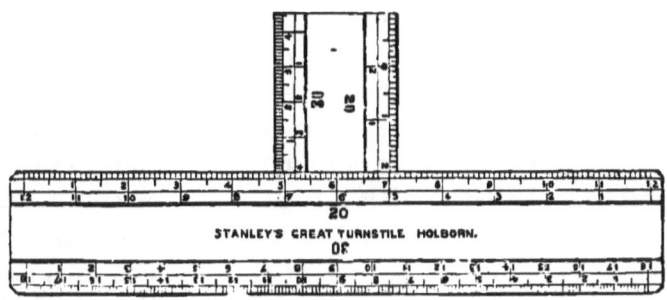

Chain Scale and Offset.

CHAIN SCALES are used by the civil engineer and land
surveyor, for plotting plans or sections of land, surveys,
railways, etc. They are universally made flat on one
side, and with the upper or bevelled edges only divided.
The divisions are marked at equal distances along
the entire scale, or what is termed *fully divided*. A
figure placed at every tenth or principal division,
represents the number of chains, and each of the subdivisions ten links.

There are six chain scales in constant use by sur-
veyors, designated by the number of divisions which
they contain per inch; these are 10, 20, 30, 40, 50, 60.

The lower scale in the illustration above represents
a 20, one-third of the actual size—that is, the scale

would contain 20 divisions to the inch, and be used in practice for a scale of two chains to the inch.

It is found very convenient to have on the opposite edge of the scale to that on which the chain scale is divided a feet scale, or what is termed *a feet equal scale*. This is a scale in which 66 divisions on the feet edge measure the same distance as 100 divisions on the opposite or chain edge. This enables the scale to be used for taking off dimensions in feet from a plan which has been plotted in links, from measurements taken by the Gunter's chain, or *vice versâ*, for bringing feet into links; the proportions of the divisions of these scales being the same as the links of the Gunter's chain in relation to feet—that is, the Gunter's chain is 66 feet long, and is divided into 100 parts or links.

Although the six scales mentioned are almost universally used for plans of estates, parishes, railways, etc., topographical surveys of Great Britain are often plotted to other scales, as 70, 80, 90, 100, $\frac{1}{2500}$th of actual size, 6 in. to the mile, 1 in. to the mile. The divisions of the metre, which are in some proportion to full size, as $\frac{1}{2000}$, $\frac{1}{3000}$, $\frac{1}{5000}$, etc., are much used in the colonies and many foreign countries besides France. These scales, when used for measuring upon maps, should be divided shorter than the true scale, to allow for the shrinkage of the map after printing; they are then technically called *shrunk scales;* the shrinkage is about a seventh of an inch to the foot.

An offset scale, which is that shown erect in the upper part of the illustration, is always used with the chain scale. It is of similar section and division to the scale, but generally only two inches long. The ends of the offset are made perfectly square, so that when slid along the edge of the scale it acts as a set square. The divisions upon the offset read from the extreme ends, thereby a perpendicular of a given distance may be set off from any part of the edge of the scale.

Many civil engineers, particularly for working sections, prefer an offset three to six inches long, sometimes with the outer line of figures reading upwards and downwards from the centre. In this instance the scale upon which the offset is used has to be placed the distance of the centre below the base or datum line, from which the offsets are to be worked. Another plan of working offsets is by means of a set square, upon which is fixed two clamps, that will hold an ordinary plotting scale parallel to the edge of the set square, and thus enable any scale to be used as a long offset.

In using the scale and offset, it will sometimes be found convenient to place a leaden weight upon the scale, to secure it from movement by the pressure of the offset against its edge. The quantities which are taken from the field book should be marked off from the scale upon the drawing with a pricker.

For the base line of a survey of considerable extent, the edge of the steel or other straight-edge should be divided at every foot with a clear line, so that it will only be necessary to add one dimension from the scale to the principal quantity taken by the straight-edge. If the straight-edge be placed about an inch below the intended line to be measured, the feet may be squared up with the offset scale.

For plotting surveys of 1, 2, or 3 chains to the inch, 18-inch scales are preferable to 12-inch. The 1-chain scale should be divided with a subdivision to read off every five links. To divide the tenth of an inch into ten parts by the eye, as is frequently attempted, is a very doubtful possibility.

There are many schemes answering the same purposes as the scale and offset, but none practically so good.

ARCHITECTURAL and MECHANICAL SCALES are duodecimally divided, the divisions representing feet and inches. The one illustrated below is what is termed *fully divided*,—that is, closely divided throughout. In the example here given, the figures read from left to right and from right to left. Another method of figuring is to make the inner line to read from the centre to right and left. This latter manner of reading

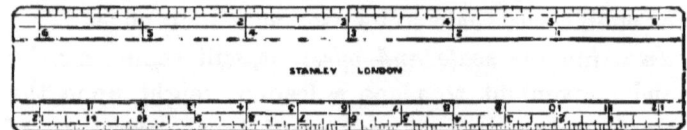

Fully-divided Scale.

enables the scale to make bi-sections, or set off equal quantities from each side of a line, a frequent convenience in every description of geometrical drawing.

Fully-divided scales are mostly used by civil engineers, as they match in every way with the chain scales; practically they are the best kind of scale, from the convenience of the feet and inches of a dimension being read off upon the same part of the scale by one observation. They are seldom used by the architectural or mechanical draughtsman.

It may be observed with duodecimally divided scales, that the subdivisions should differ according to the size or proportion of the scale; the reverse of this is a very common fault in scales as they are ordinarily made.

The divisions upon the small scales are often much too close, with the foot divisions subdivided to represent inches. No scale for either architectural or mechanical drawing should have any divisions closer than 48 to the inch. Thus, it is better to have the $\frac{1}{16}$ scale once subdivided, every division representing six inches; the $\frac{1}{8}$-inch scale subdivided in four or six, each division representing three inches or two inches; the $\frac{1}{4}$, $\frac{1}{2}$, and $\frac{3}{4}$-inch each into 12 for inches; the 1-inch and $1\frac{1}{2}$-inch each into 24 for half-inches; the 3-inch into 48 for quarter-inches, and the intermediate scales in similar proportions. Wider spaces than these are uncertain—closer divisions are very perplexing: moderately careful observations will subdivide the interspaces with sufficient accuracy and greater certainty than by marking off from a confusion of lines.

OPEN DIVIDED SCALES have the lines to represent feet carried entirely along the scale, and the subdivisions, which represent inches or parts, placed at one or both ends, outside of the feet divisions. Thus, in using these scales to set off a dimension, the inches have to be set off from near one end, reading backwards from the first division of the feet, and the number of feet are read off along the scale afterwards.

Open divided scales are generally made with two scales upon one edge; sometimes they are made with one scale only, reading to right and left. They are also divided and figured differently, as is shown in the illustrations on the opposite page, each of the three representations being one manner.

These scales are more generally made of oval than of flat section. The first figure on the next page shows the most useful; this, as generally made, contains the $\frac{1}{8}$, $\frac{1}{4}$, $\frac{1}{2}$, 1, $\frac{3}{8}$, $\frac{3}{4}$, $1\frac{1}{2}$ and 3-inch scales.

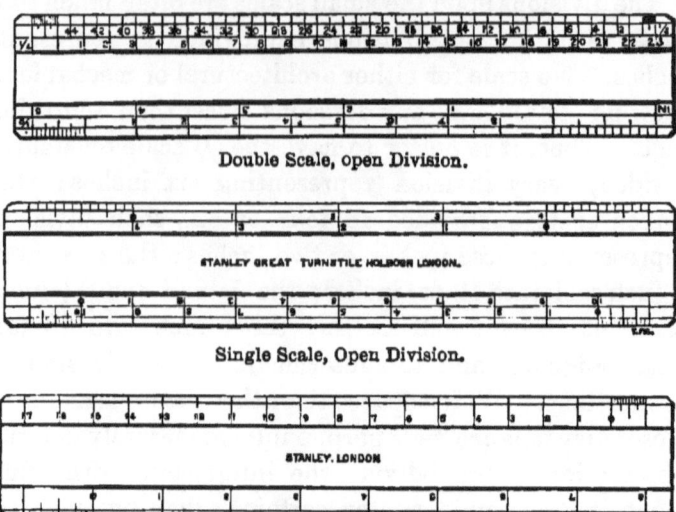

Double Scale, open Division.

Single Scale, Open Division.

Single Scale, Single Figuring.

A very clear scale for mechanical drawing, for the open scale, from 1½ to 6 inch (half size), is produced

Single Reading, Full Figuring.

by having a simple reading from left to right, fully divided throughout, similar to a reduced ordinary measuring rule; the feet are marked with a rather large figure, and all the inches, 1 to 11, separately marked with a small figure. The scale made in this manner comes out remarkably distinct; if the ¾ and inch scales are desired to match, they may be figured for the inches 3, 6, 9, only. The engraved specimen of a piece of a 1½-inch scale, full size, is shown above.

Open divided scales appear plainer at first sight than fully divided ones; they cannot really be so expeditious in use, although after considerable practice with the one kind it is difficult to adopt the other.

Architects' Universal Scale.

Builders' Universal Scale.

UNIVERSAL SCALES, although very generally employed, are not to be recommended. They contain a confusion of scales seldom required, which tend to cramp and perplex the useful ones. The foregoing illustrations represent the upper sides of the two kinds most popular. In the first illustration, the ordinary open divided scales are placed upon the edges, and there are also four or more scales along the centre of each side of the scale. These can only be used by taking off the required quantities with dividers; therefore they are of little practical value. In the second illustration, which is that technically called a *Builders' Scale*, the divisions all read to the edge, each edge containing three or four scales, which read from left to right—very confusedly.

The scales already described embrace all those mostly used by architectural and mechanical draughtsmen; a few others are occasionally made for especial purpose,

P

or fancy. The one illustrated below appears to the author a useful one of this class; it is called a *Joist and Brickwork Scale*: it is used to mark off joists or

Joist and Brickwork Scale.

rafters, towards one end, and lengths of bricks towards the other, to a given scale; allowing 12 inches space, and 2½ inches timber for the joists, and 9 and 4½ inches for the bricks. The ordinary scale is placed on the opposite edge, as is shown in the illustration.

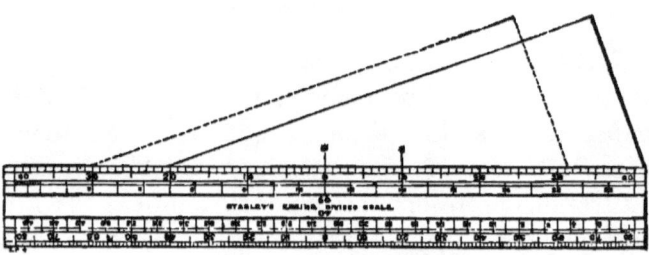

Marquois' Scale and Triangle.

MARQUOIS' SCALES are used for military drawing only, for which purpose they possess some essential qualities. They are very portable, have much greater solidity than ordinary scales, and their peculiar form adapts them to supply the place of the square, set square, straight-edge, and parallel rule, for limited sized drawings.

The set of Marquois' scales consists of two scales of equal width, one only of which is illustrated above, and a set square, or *triangle,* as it is termed. The whole three pieces are generally made of stout box-wood of about a quarter of an inch in thickness.

The triangle has the length of two of its sides in the proportion of 3 to 1, the longest side, or hypothenuse, being three times the length of the base; the remaining side, which is at right angles to the base, is bevelled off so as to present a rather thin ruling edge. In the centre of the longest side is a line with a star, called an index; this line reads into the scale when the triangle is placed against it, as shown in the illustration.

The two similar scales have in all eight pairs of scales divided upon them—that is, two upon each edge of the four sides. Each of the pairs of scales gives one of the following number of divisions per inch—20, 25, 30, 35, 40, 45, 50, and 60. They are here termed *pairs* of scales, because, although differently divided, the divisions represent, in use, similar quantities, the difference being in the system of notation.

Of the pair of scales—the inner line of division is called the *natural scale,* the outer line nearest the edge is called the *artificial scale.*

The *natural scale* is decimally divided in the manner of the open divided scales described in this chapter, the tens being carried along the scale, and the units placed at the ends only. As the divisions do not read to the edge, quantities can only be taken from these scales with a pair of dividers.

The *artificial scale* is a fair illustration of the peculiar artificial system of scales in common use by the profession of the last century. The divisions upon this scale, which read from the centre each way, are made three times as wide apart as their nominal indication; for instance, the 20, instead of being 20 to the inch, is divided into 20 in three inches; thus it agrees with the proportions of the sides of the triangle, with which it is intended to be used in a peculiar artificial manner, which will best be described by an example.

To measure a distance from a given line with the artificial scales, the bevelled edge of the triangle is placed against the line, the scale is then placed against the longest side of the triangle, with the 0 upon the scale reading into the index line in the centre of the angle. If the scale be held firmly in this position, the triangle may be slid along it until the index is brought to the required quantity, according to the reading of the artificial scale. A line now drawn along by the bevelled edge will give the actual quantity.

It will be observed, by this system, that the triangle is moved along the scale three times the distance that the bevelled edge recedes from the first line, the divisions of the scale being also three times wider than the natural scale. This is clearly shown by the dotted lines in the illustration, which indicate the position of the instrument before movement.

The advantages claimed for this system are, that the lines being three times as wide apart as indicated, distances may be better read off, and that parallel lines may be drawn at equal distances apart with less error. Of course it will suggest itself to any draughtsman using it, that if the scale were a natural scale, and the same distance of parallels were required, it would only be necessary to move the set square three divisions at a time, instead of one at a time, to produce the same effect.

It appears somewhat curious that this antiquated system of scales should be still retained as a portion of military education, when such artificial systems have been for many years abandoned by the architectural and engineering professions. These scales have only one merit, solidity; this is of importance to the military officer; but on the other hand they are deficient in the constant convenience of edge reading, and require every dimension to be taken with the dividers, or, what is more tedious, by the artificial system.

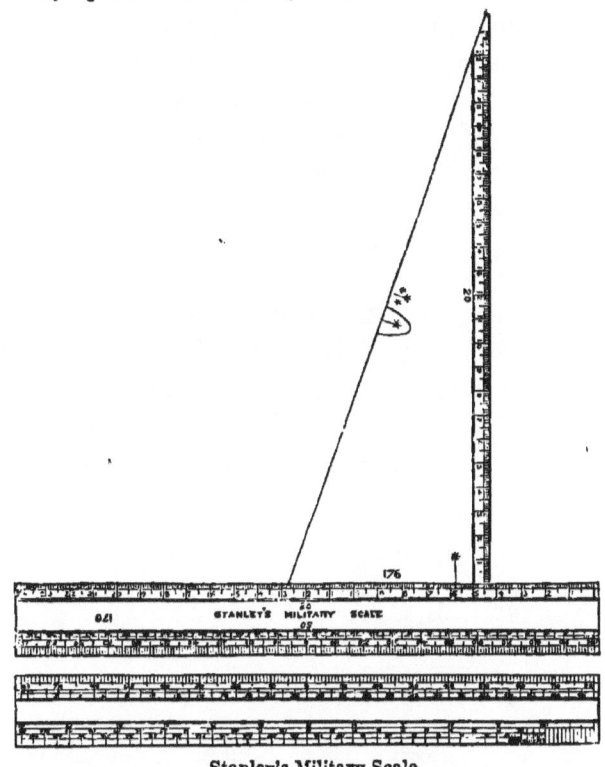

Stanley's Military Scale.

The author would suggest that the scales might be conveniently remodelled, retaining all their merits by making three of the edges *bevelled*, to contain the scales in most frequent use—10, 20, and 30; and that the triangle should be divided as an offset to one of these—say the 20. Four of the artificial scales might then be placed on the flat surfaces at the back, as usual, if required; but perhaps three would be *quite* sufficient on the back, and one on the face; thus we should have 15, 40, 50, and 60; in the place of the fourth on the back, a scale of feet and inches might be divided, which would be constantly useful.

The above remarks appeared in the original edition, since which the author has been asked to arrange a suitable scale for military men, which he has done; it embraces the improvements suggested, and is now adopted at our military colleges. It is shown in the engraving upon the last page.

Military scales are generally placed in a slide lid box. Formerly at Addiscombe College they were fitted under the tray of the case of instruments; this obviated the necessity for two boxes, one for scales and the other for instruments, and appears to the writer to be the better mode of keeping them for military men.

CHAPTER XXIX.

MATHEMATICAL LINES OR SCALES PRINCIPALLY DEDUCED
FROM GEOMETRICAL FIGURES—GUNTER'S SCALE—
PLAIN SCALE—SECTOR—SLIDE RULE, ETC.

MATHEMATICAL LINES are scales of proportions which
serve commonly for geometrical calculation or illustration, and are, generally speaking, more valuable for
educational than for practical purposes.

To describe them fully would occupy space beyond
our limit, and would be superfluous, as the subject has
been completely studied and written upon a century
past, and detailed and retailed in every diversity of
form in stereotyped matter.

For practical purposes, instead of very fallible scales
of proportions, derived from geometrical form, we have
very exact tables of logarithms and algebraic formulæ,
endowed with superior powers of solving the abstruse
questions of navigation, astronomy, and dynamics. It
would, however, appear an omission if some notice
were not taken of the lines which frequently occur on
some of these mathematical scales—as on the Gunter
scale, used for navigation, the plain scale, some kinds
of rectangular protractors, the sector, etc.

The DIAGONAL SCALE, now nearly obsolete, was formerly one of the most universal mathematical scales;
it is still placed, as a matter of form, on the protractor
supplied with most cases of mathematical drawing instruments. Theoretically, it is a very ingenious scale;
practically, it is an almost useless one—the only pur-

pose to which it is now applied being to a scale for the beam-compasses, which to be of any use, should be divided upon metal.

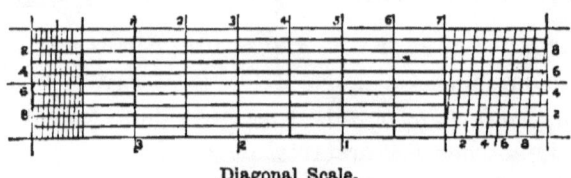

Diagonal Scale.

The purpose of the diagonal scale is to divide any given quantity into some number of equal parts, mostly 100 parts.

The illustration above represents a portion of a half and quarter-inch diagonal scale, which is constructed as follows. Eleven lines are drawn *longitudinally*, upon the scales, at equal distances apart; the scale is then divided into quarter inches, the divisions crossing all the lines. *One-half inch* at the right-hand end, and *one-quarter inch* at the left-hand end of the scale, are each divided into ten parts, the divisions being dotted on the top and bottom lines only. These dots are then united by lines from top to bottom, the distance of one space out of perpendicular. Thus the first dot on the top line is united to the second on the bottom line, the second on the top to the third on the bottom, the third to the fourth, the fourth to the fifth, and so on,—the whole united dots forming oblique parallels.

The principle of this system is that the top and bottom lines being each subdivided into ten parts, the diagonals crossing from one subdivision to another, in advance; this subdivision is thereby lengthened out, and again subdivided by the longitudinal lines into ten parts, making the total division of the nominal quantity into 100 parts.

Thus, the crossing of every diagonal line advances on the half-inch diagonal the two-hundredth of an inch from one longitudinal line to another, and on the quarter-inch diagonal the four-hundredth of an inch.

By this system to measure any tenth of the half-inch, taking the right-hand diagonal scale of the illustration, the bottom line would be taken; to measure any number of hundredths with one as unit, as $\frac{11}{100}$, $\frac{21}{100}$, $\frac{31}{100}$, etc., the second line from the bottom would be taken; to measure any number of hundredths with two as unit, the third line would be used, and so on for the other lines, calculating the number of tens by the number shown in the bottom division, and the number of units by the longitudinal lines, which count in this scale upwards from the bottom. The half-inches and quarter-inches being ruled through all the longitudinal lines, any number of the dimensions may be set off on any line, according with the position of the required fractional part.

Although half and quarter inch is given in this example divided in ten, any other quantity may be divided into any number of equal parts, in similar manner.

It must, however, be admitted that this excellent theory is rendered very useless in practice for the minute divisions to which it is sometimes applied, from the fact that the scale must be hand-divided, and is consequently incorrect. The truth of this remark may be readily observed, on any ordinary rectangular pro-tractor, if a quantity be taken from one end of the double diagonal scale, and the same quantity be read off on the other end of the same scale, by double reading.

With regard to the other lines upon mathematical scales, the line of numbers marked N is a line of geometrical proportion, by means of which operations in multiplication and division may be performed with greater trouble and less accuracy than by figures.

Cho.

Line of Chords.

The LINE OF CHORDS, marked "Cho.", is used to protract angles, which operation it will perform with a little more trouble, and, unless very accurately divided, with less exactness than by the protractor, to be described in the next chapter. It is, nevertheless, perhaps the most useful of what are termed mathematical lines. It may be practically used for portability, as, for instance, upon the small scale of equal parts commonly packed in a case with Napier or pillar compasses. In this instance it is found particularly convenient, as the chord line does not occupy the edge space which is required for the ordinary scales.

The chord line is used to protract an angle in the following manner. A pair of dividers are first set by the chord line to the space from the 0 to 60, and with this distance as a radius an arc is described of sufficient size to contain the required angle. If the dividers be again applied to the chord line, and the distance of the number of degrees required to be taken from 0, and pricked off upon the arc, these degrees will, if united with the centre from which the arc was struck, give the required angle.

The *line of rhumbs,* marked R. H. on mathematical scales, may be used to lay down the angle of a ship's course upon a chart. It is an enlarged line of chords, being the chord of the thirty-second part of the circle, set out to correspond with points of the mariner's compass instead of to degrees. It is used in exactly the same manner as the line of chords.

The *lines of tangents*, marked Ta—*semi-tangents*, S. T.—*secants*, Sec.—and *sines*, S., may be used for the stereographic and orthographic projection of the sphere, etc.

The *lines of latitude*, La.—*longitude*, Lon.—*hours*, Ho., etc., are used in the construction of sundials, therefore are in no manner a part of our subject.

The above-described lines also occur upon the sector, but they are worked out upon a different system.

The SECTOR is a kind of twofold rule, commonly supplied with a case of mathematical instruments, as a kind of established ornament, in most instances to be practically used only as a kind of bevel to erect angles, in the manner described in the previous chapter in treating of the isograph. It is, nevertheless, an ingenious instrument, of which an adequate description would occupy many pages that could scarcely be better written than the descriptions given and extracted by numerous authors.* It is only on account of its very limited practical use that merely a sketch is given here.

The theory of the construction of the sector is—that the arms opening like radii from a centre, they will thus form any angle, and that by dividing the arms, any rate of proportion may be obtained by applying a pair of dividers to the divisions from arm to arm. This may be described in the simple lines marked Pol., for

* Cunn's "Treatise on the Sector," 1729; Robertson's "Treatise on Mathematical Instruments," 1775; Adam's "Essays," 1791; Sims's "Drawing Instruments," 1845.

polygons, upon the sector, which may be used to divide
the circumference of a circle into any number of equal
parts, from 4 to 12, thus :—

If a given radius be taken with a pair of dividers,
and the sector be opened, so that each point of the
divider fall upon each line marked 6 on the line of
polygons, and the sector is kept in this position, the
distance of 5 to 5 will divide the same circle into 5,
the 7 to 7, into 7, and so on. The six is taken above
because the radius of a circle divides the circumference into 6.

By this example it will be seen that if the sector be
set to one known side of a polygon, it will give the
side of other polygons that may be described in the
same circle, the sides of polygons following their own
peculiar geometrical proportion. In the same manner
the sector will give the diminishing proportions of
chords, sines, tangents, secants, etc.

SLIDE RULES.—Another system of calculating scales
is the very ingenious sliding scales attached commonly
to mechanical engineers and carpenters' rules. These
perform many useful operations in arithmetic, with
greater facility and convenience than the sector performs its peculiar problems, as no separate instrument
is required to be used with them. Description here
would be out of place in a work on drawing instruments
and would occupy much space. Every mechanic may
be supplied, at a very trifling cost, with a book containing full instructions for the use of the slide, at the
time he purchases the rule.* The sliding system of

---

* A comprehensive description of the slide rule is given in
Dr. Rees's "Cyclopædia." See also "Two Treatises on the
Slide," by W. H. Bayley.

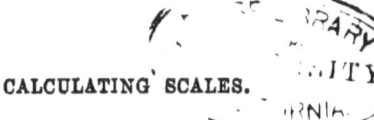

scales, now nearly obsolete, was formerly used for numerous calculations, especial scales being made for the various requirements; hence we had excise rules of various kinds, timber rules, etc.

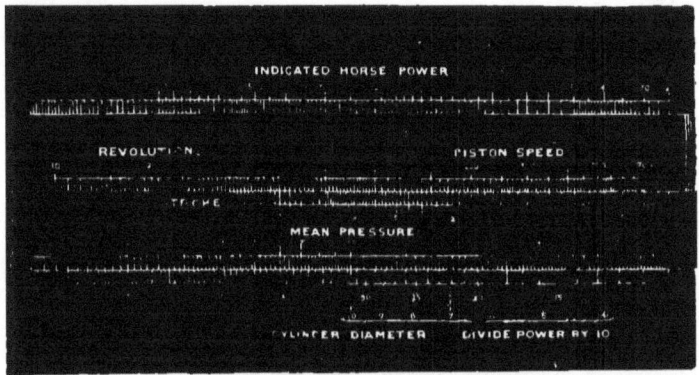

Hudson's Horse-Power Computing Scale.

HUDSON'S HORSE-POWER COMPUTING SCALE. — One of the most ingenious and simple rules of this class, for calculating at a glance all proportions of steam engines, has lately been introduced by Mr. Hudson, C.E., which he terms a *Horse-power Computing Scale.* It will be found a most valuable instrument to every mechanical draughtsman. The above engraving is from a photograph of the instrument, on a basis three-fourths the actual size. It has two slides, both moveable in either direction. It is made in cardboard, and engraved from engine-cut steel plates. It may be carried in the pocket, or in a light leather case. The instructions for its use will give in the most concise manner the purposes to which it may be applied.

Instructions for Use.
To find Power of Engine.—Set the "Mean Pressure"

on lower slide against the "Cylinder Diameter," and retaining it in this position, bring the "Revolutions" on upper slide, opposite the "Stroke" (or the "Piston Speed" opposite the small arrow); the large arrow will then point to the "Power."

To find Size of Engine.—Set the large arrow on upper slide against the "Power" desired, and retaining it in this position, bring the "Stroke" on lower slide opposite the "Revolutions" (or the small arrow opposite the "Piston Speed"); the "Mean Pressure" will then be against the size of cylinder necessary.

To find Piston Speed.—Set the "Stroke" against the "Revolutions" and the small arrow will point to the "Piston Speed."

Mean Pressure Scale (on back) shows what proportion of initial pressure is realised as mean pressure, with cut-off at different percentages of stroke. For a Compound Engine (with two or more cylinders), set the number on "Mean Pressure" slide representing the percentage of cut-off in smallest cylinder, against the diameter of that cylinder, and use the number then found opposite diameter of largest cylinder as the cut-off for the latter working by itself, to give power equal to that from the compound cylinders.

Ratio of Compound Cylinders.—Set the 1 on "Mean Pressure" slide against the diameter of L.P. cylinder, the ratio will then be opposite that of H.P. cylinder.

To the intelligent mechanic the sliding scales have undoubted value. The writer would suggest that their use might be taught with advantage in our public schools, as a most convenient part of a mechanic's education, particularly as the gauge points, as they are termed, require more effort of memory to retain than reflective power to execute the dependent operations.

We have lately published a very concise and lucid description of the slide rule, accompanied by a cardboard moveable diagram, by Mr. Charles Hoare, C.E., in Weale's popular series of scientific books. Any person who has not entered upon the matter will be astonished at the complicated calculations that may be effected by a few simple movements. The same kind of rule from which the diagram mentioned above is taken, is much used for a pocket rule on the Continent, where the slide rule is better understood. It is made of boxwood, very light and durable; the same, of course, is occasionally made in London and elsewhere; but not being much understood, the demand is not sufficient for it to be frequently met with.

CHAPTER XXX.

INSTRUMENTS TO DIVIDE THE CIRCLE—GENERAL DE-
SCRIPTION—VERNIER READINGS, ETC.—PROTRAC-
TORS OF VARIOUS KINDS—STATION POINTER, ETC.

EXPERIMENT has proved that it is one of the most
difficult of mechanical operations to accurately divide
the circumference of the circle into equal parts. It is,
at the same time, of the greatest importance that the
circle should be so divided, for the many scientific purposes to which it is applied. To astronomy and navigation, it is the rule by which the earth and the visible
heavens are measured. It has taken many years of
conscientious labour, devoted by our most scientific
workmen, to produce the machines by which we have
attained the passable accuracy with which our instruments are now divided. The entire problem remains
an unaccomplished possibility. It would diverge too
far from our subject, and swell these pages beyond
their intended limit, to name the workers and to tell
what they have done. It is only mentioned as a note
in passing, the nearly attained being all-sufficient for
the simple processes connected with the divided circle
when applied to drawing instruments.

The circle, large or small, for scientific purposes, is
uniformly divided into 360 equal parts, which are
termed degrees; each of these degrees is divided, or
presumed to be divided, into sixty parts or minutes,
and each minute into sixty seconds. In practice,
particularly for drawing purposes, the circle is seldom

divided closer than to half-degrees ; when the further
division into minutes is required, it is accomplished by
what is termed a *vernier reading*. This is a very simple
and exact method of subdividing, that owes its high
scientific value to a simple theorem in natural optics
—That the eye can readily observe whether a line be
continuous or broken, although it can in no way accurately measure the distance of any separation. It will
be necessary to fully describe the vernier, as it is so
constantly used wherever exact division of the circle
is required.

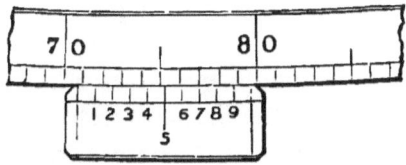

The VERNIER SCALE is a short scale attached to the
divided edge, or what is technically termed the *limb*
of the instrument, in such a manner that it will slide
evenly upon or around it, and that the divisions upon
the vernier will form continuous lines when opposite to
the divisions upon the limb.

The vernier being used to subdivide the divisions of
the limb, it is divided, for this purpose, into as many
spaces as the subdivisions required. These spaces correspond with the lines upon the instrument within one
division in the quantity; thus—To divide half-degrees
into minutes—that is, into 30 equal parts—the vernier
would have 30 spaces between the divisions upon it, but
the 30 would be divided either in a distance equal to
that occupied by 29 or 31 of the spaces upon the limb.
In practice, the 29 is uniformly adopted. In reading

Q

the vernier, the fractional quantity is taken at the
division where it becomes coincident with a division
upon the limb,—that is, where the line appears continuous.   This will be better explained by reference to
the illustration, which, for clearness, is shown dividing
the degree into ten equal parts.   In this the space of
nine degrees of the limb are taken off upon the vernier,
and this space is divided into ten equal parts.   To
read it in the position here shown, the division on the
limb which is nearest behind the first division of the
vernier is taken, which, in the figure, is 70.   Now to
know how many tenths of a degree it is past the 70,
we refer to the vernier; here we see that the seventh
division of the vernier forms a straight line with one of
the divisions upon the limb, therefore the reading in
the present position is 70·7 degrees, or 70 degrees and
seven-tenths of another degree.

The 7 of the vernier in the illustration reads into
the 77 of the limb; this may cause a little uncertainty
as to which the fractional 7 is taken from.   It may be
well to repeat that the fractional reading is taken from
the *vernier only*, where one of the lines appears continuous with any one of the lines on the limb, without
regard to which one, and the quantity itself from the
first division of the vernier.

It will be obvious that the vernier is equally applicable to the straight line as to the circle, although
seldom applied to drawing instruments, except occasionally to the beam compasses.

Before describing the circle-dividing instruments to
which the vernier is applied, it will be necessary to
describe the simple protractor used for marking off the
degrees by direct observation.

The PLAIN CIRCULAR PROTRACTOR, of which the following is an illustration, is generally made of brass or electrum. The circumference is divided into degrees,

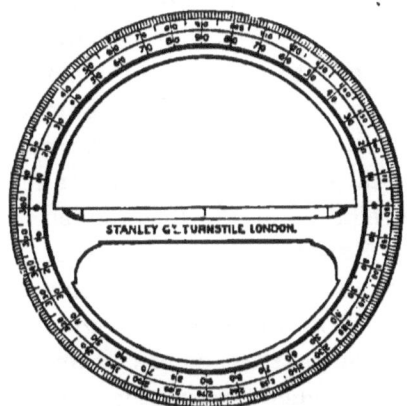

Circular Protractor.

which are generally subdivided into half-degrees. The extreme edge is bevelled down very thin, so that the needle point, used in marking off the degrees, may feel the divisions. The bar that crosses the inner space is made so that one side of it is a true line through the centre of the protractor. This line, if continued, would read into 0° and 180°. The outer edge of a line in the centre of this bar is the centre of the circle from which all the degrees are set off.

It may be observed that it will be found difficult to place the outer edge of this line over a point; to remedy this, a small semicircle may be taken out of the edge of the bar; the point would then have to be set by judgment, which, although apparently less exact, is much more convenient.

This being a simple protractor, it is seldom used with a vernier scale. The only vernier applicable is a

loose vernier with an arm, which is not shown in the
illustration, as it is seldom applied.

*In using the plain circular protractor for setting off
an angle from a point in a given line,* it is necessary
to place the bar so that it half covers the line, which
should read from 0 to 180 in the division, and that the
point appears in the centre; if the number of degrees
be then pricked off, and the protractor removed, a line
may be drawn through the point and the centre,
which is the required angle to the given line.

The PLAIN SEMICIRCULAR PROTRACTOR is similar to
that just described, except that it only reads 180 or
200 degrees. Those reading 200 degrees are most convenient; the centre then comes upon the top of the
bar, as in the circular protractor. The semicircular
protractor is quite sufficient for architectural and
mechanical purposes, and much more convenient than
the circular, as it may rest on the edge of the tee-
square when an angle is required to be erected from a
horizontal line.

Protractors used for exact purposes, as plotting surveys, etc., have some means of extending the degrees
beyond the edge of the protractor, also the more accurate manner of reading by means of a sliding vernier.

The figure on the next page illustrates a *Circular Protractor, with Vernier and Arm.* The same construction
is also commonly applied to *semicircular* protractors.
The vernier has been already described. The arm consists of a piece of metal jointed round the centre, and
extending beyond the circumference of the protractor,
one side of it forming a radial line from the centre.
The part of the arm which extends beyond the circumference lies in close contact with the paper, so that

a line drawn along by the side of it will exactly correspond with the reading of the vernier.

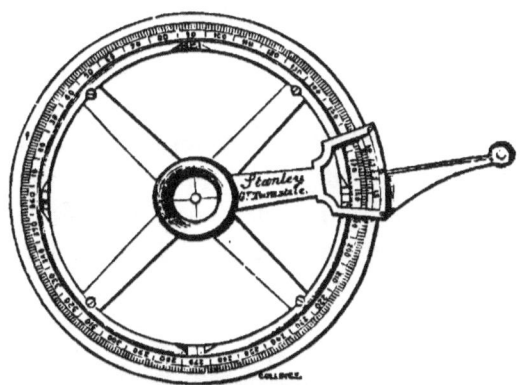

Circular Protractor, with Vernier and Arm.

It is, however, found difficult to draw a line by the side of the arm to exactly correspond with its edge; it will be found better to make the arm thin, so that it may have a slight spring, and to have a sharp point fixed to the end of the arm in a line with its ruling-edge; this, by gentle pressure, may be caused to puncture a faint hole upon the drawing, which will give the true position of the angle, according to the reading of the vernier. The point may be united by a line to the centre, after the protractor is removed.

The centre of this protractor answers best if made of glass, with fine lines across it; the glass, being transparent, will allow the centre to be placed over the point with facility. Upon the suggestion of the late Col. Strange, Inspector of Scientific Instruments for India, the author has made no true centre, but a small circle around it, with the cross-lines leading up to the circle. The centre in this manner may be readily observed, without the risk of being deceived by its being under one of the lines instead of in the crossing. The eye will detect a point in the centre of the circle with great precision.

The bars which support the centre of protractors that read with a vernier are not made in a line through the centre as in the plain protractor, but the line from which an angle is to be set off on the drawing is placed under two lines, which are drawn down on two opposite bevels, on the *inside* of the circumference of the instrument.

Folding Arm Protractor.

The FOLDING ARM PROTRACTOR, above illustrated, is the most perfect instrument for accurate plotting. The construction of the divided circle, vernier, and centre is similar to the last described. The folding arms and apparatus connected with them are fixed upon a kind of frame, the whole of which moves upon an axis in the centre of the instrument. Upon the frame are fixed two verniers, which read into the opposite sides of the divided circles, and serve to correct any possible in-accuracy in the construction or division of the instrument. From the centre of the instrument, at right angles to the verniers, a portion of the frame is brought to the extreme circumference; this carries what is termed a *clamp and tangent motion*, which consists of means of fixing the verniers roughly in any position, and afterwards delicately adjusting them by a screw.

Upon the ends of the frame, near the verniers, the folding arms are jointed upon adjustable pivot centres, which admit of the arms folding back over the centre of the instrument, to render it portable when out of use. The folding arms are light triangular frames of metal, which are supported from the surface of the drawing by small springs. The extreme end of each of the arms carries a point, which, by light pressure over it, leaves a puncture upon the drawing exactly corresponding with the reading of the instrument.

The great accuracy of this protractor, in comparison with any other, is derived from the arms being opposite, and the points which they puncture at a considerable distance; thus the angle is set off correctly, independently of the centre; at the same time it is a test for the accuracy of the instrument, or the truth of its position upon the drawing; because if the centre be placed over a straight line, and the points fall exactly into this line, the instrument will be in perfect adjustment; or, in using it, if a line drawn through the punctures made by the points pass exactly through the centre over which the instrument was intended to be placed, it will show if it were placed accurately.

A very scientific workman, who will not allow me to

mention his name, has invented a protractor, which I consider to have great merit. The protractor is made in two parts, or rings, concentric the one to the other. The inner part carries the cross bar only, the outer

Inclination Protractor.

can therefore be placed at any angle to the outer cir-cumference, which is the protractor proper. The top surface of the outer circle is divided, and reads upon the inner circle by a vernier. Two lines only are drawn down to the surface on the outer edge of the outer circle. The protractor is first set to any re-quired angle; then being placed upon a line, which cuts the two lines of the outer circle, the bar will direct a clear line through this, at any point on the first line desired.

CAPTAIN DOUGLASS' REFLECTING PROTRACTOR.—This is a protractor with vernier and arm carrying a pair of reflectors similar to the sextant, the principle of which is described in the Optical Compasses at page 136. This instrument can be held in the hand, and true angles taken as with the sextant, and the angles can be at once drawn by the instrument. It is very accurate for sketching upon the ground, but would scarcely be

used by any one who understood the use of the field
book.

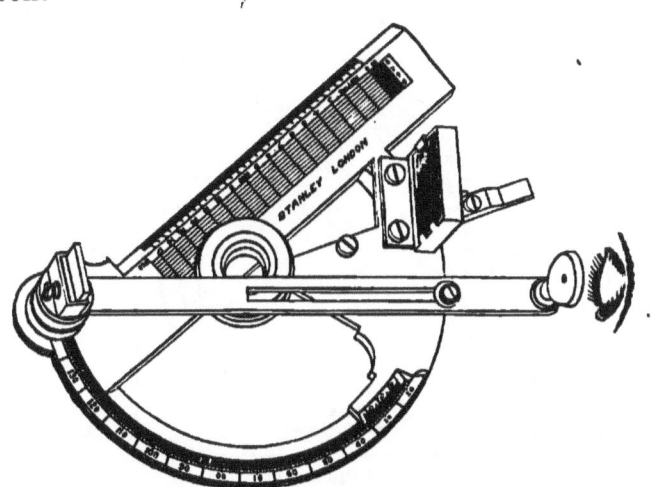

Captain Douglass' Reflecting Protractor.

The CARD PROTRACTOR, of which the following engraving is an illustration, is a very correct instrument,
that may be purchased at a moderate price. The
divisions are printed from an engine-divided steel or
copper plate, on ivory card, the circle being generally
twelve inches diameter. The divisions are made to read
inwards of the circumference instead of outwards, as
in other protractors, the centre space of the card being
entirely cut away. This makes it appear somewhat
tedious to use.

*If degrees are required to be set off from a point on
a given line,* to bring this point into the centre of the
protractor, it is necessary to erect a line at right angles
over the point, and to make this line read into the 90°,
at the same time as the first, or given line, reads into
the 0° and 180°. If the protractor be placed in this

position and held by leaden weights any number of angles may be set off from the centre formed by the meeting of the lines. Although the setting is rather

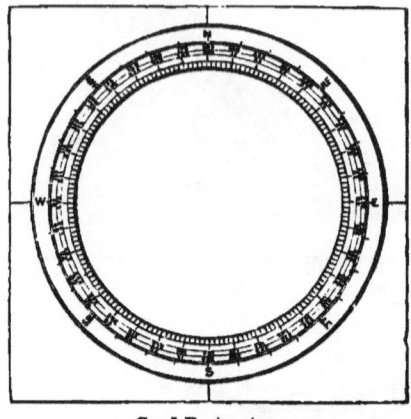

Card Protractor.

tedious, from the trouble of finding the centre, there is some compensation in being able to draw in many angles at once, by means of a small straight-edge, without the necessity of removing the protractor.

HORN PROTRACTORS in form and division resemble the plain protractors described; they are generally made semicircular. For many purposes they are peculiarly convenient, from the transparency of the horn, which allows the whole of the drawing beneath to be plainly seen. For exact purposes the unequal expansion and contraction of the horn render them too inaccurate. They also cockle up and become difficult to use. The best manner of keeping the horn pro-tractor passably straight is to place it in a book or under a weight when out of use.

The writer has made a protractor of nine inches diameter, similar in appearance and transparency to horn, the material of which is what is termed horn paper, a paper prepared with drying oils; it is, perhaps, rather less transparent than horn, in other respects it is better; it does not contract unequally to derange the divisions, and it keeps perfectly flat. Another advantage of this material is that protractors of much larger sizes may be made of it than could be made of horn.

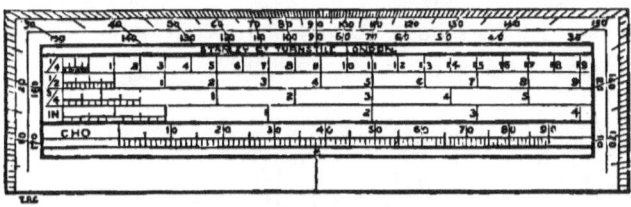

Rectangular Protractor.

RECTANGULAR PROTRACTORS are such as are generally supplied with cases of drawing instruments, and are either made of boxwood or ivory. One of the ordinary descriptions is illustrated above. They are generally covered with scales of the kind already described in a previous chapter,—open-divided, diagonal scale, etc. The scales are of very little use for practical purposes, as they are too short, and do not read to the edge.

The protracted line, which it divides into degrees, is carried along the edge of one side and of both ends; the figures are in two lines, and read from 0 to 180, proceeding from the base or centre line both to the right and left, thus bringing the two 90° to the centre of the protractor. Upon the base a line is drawn, the outer edge of which is the centre from which all the angles are set off.

The inequality of the degrees at the edge, and the shrinkage of the material after division, if of wood or ivory, render this protractor unfit for exact purposes ; the principal use made of it is for sketching perpendiculars, which is done by placing the centre and 90° on the horizontal line, and drawing the perpendicular or vertical line by the base ; for this purpose, however, it is less expeditious than the straight-edge and set-square. Although it is generally supplied with cases of drawing instruments, it is now seldom used except in schools, and occasionally by architects, who have little use for any kind of protractor.

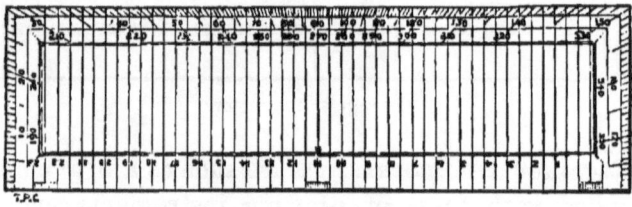

Military Protractor.

The MILITARY PROTRACTOR, employed for military sketching, is portable, and in every way sufficient for the purpose. It is uniformly made of ivory, and is similar, as regards the protracted edges, to the last described. The distinguishing feature is a series of lines across it, which are made alternately red and black; these lines are ruled up from a scale divided along the base, to which they are perfectly vertical, being parallels with the line which reads from the centre of the protractor to 90°. These lines present a peculiar advantage for making sketches of field work, the readings of which have been taken with the prismatic compass, as they are made to represent the parallel of latitude, or east to west in any part of the protractor.

To make this comprehensible, it will be necessary to show the manner in which an angle or bearing taken with the prismatic compass is transferred to the drawing.   In the first place, the paper on which the plot is to be produced is ruled entirely over with parallel lines at unequal distances, say, one inch to two inches apart; these lines represent the direction of magnetic east to west, technically of 90°, as do also the lines across the protractor.

Now, presume that a line is required in the direction of 135°, which has been taken by the prismatic compass, and which would represent the magnetic south-east, and this is to be set off on the plot from the centre of the station point, which should be marked ⊙ on the drawing.   If this point occur on one of the lines we have drawn, we should place the protractor with the centre upon the point, and the line reading through the centre up to the 90° of the protractor.   But if the station ⊙ does *not* occur on one of the parallel lines we have drawn, we place the centre of the protractor still over the station marked ⊙, and observe which of the parallel lines upon the protractor is most coincident with one of the lines upon the paper.   All lines upon the protractor are 90° and parallels, and all lines on the paper are considered as 90° and parallels, therefore either of these lines gives the true direction equally well with the actual 90° of the protractor; thus, we have only to set it to parallel position, prick off the degrees, remove the protractor, and draw the line from the station point to the mark pricked off.

Various scales of equal parts are introduced upon military protractors, according to fancy; the above description gives the distinguishing feature only.   An

elaborate and excellent description of the uses of the
military protractor in connection with the prismatic
compass is given in "Major Jackson's Course of Military
Surveying," to which we would refer the student
who intends to embrace the military profession.

The SANDHURST PROTRACTOR is a military protractor
adapted especially for topographical delineation, and is
different to many instruments of its kind in having

useful matter only upon it. It is made of boxwood,
upon which the protractor is cut, and has also one scale
at the lower edge of six inches to a mile in yards, the
tens of which are carried across to make parallels of
90°, in the manner of an ordinary military protractor.
Over the back of the protractor is a scale which gives
a standard for shading slopes of land upon topographical
maps from two to thirty-five degrees, also lines for
contour shades.   A small plummet is supplied with the
instrument, the cord of which is passed through a hole
in the centre from which the degrees are protracted.
When the protractor is held up, degrees downwards,

the cord of the plummet will pass over the degrees
and indicate the angle at which it is held; by looking
over the edge in this manner the angle of inclination
of the land may be taken as with a clinometer, or by

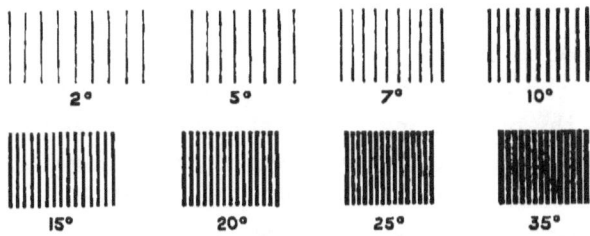

Example of Scale of Shade for Slopes.

looking along the edge (a second person reading the
plummet) angles of altitude may be taken.

As a topographical sketching instrument it is the
best that can be had at a low price.

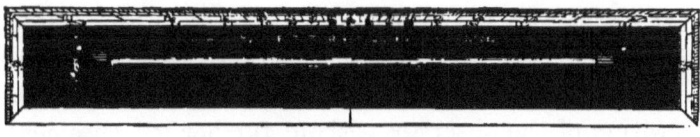

New Naval Protractor.

NAVAL PROTRACTOR, *for laying down a ship's course.*—
The writer, upon the suggestion of the requirements of
our captains in the Oriental Company's steam ships,
invented a protractor which should at the same time
form an effective parallel rule; this is illustrated above.
It consists of a metal parallel rule, the edge of which
is protracted to half degrees; the difference from other
instruments of the kind is that the bearings of the
roller are supported upon springs in such a manner
that the instrument acts as a parallel rule when moved

without weight upon it, but by a slight pressure the roller recedes into the bed, it then acts as a protractor only; thus in running up a parallel the course can be protracted off at any point.

ISOMETRICAL PROTRACTOR is constructed for isometrical perspective drawing, to give all angles from the horizontal in correct ratio. Four scales are divided upon the surface, besides the protracted edges, which give the proportions of natutal scale to horizontal, and to vertical, and to the diagonal of 45 degrees. By these scales, ovals that represent the circle in isometrical perspective may be drawn through eight or ten points, which are proportional radii upon the scales.

The STATION POINTER is a kind of double arm protractor, with which two angles relative to a base may be taken simultaneously. It may be conveniently used in plotting or sketching new countries, or for taking stations inaccessible to measurement, and is used generally for coast surveying. With this instrument, the relative angular position of three known objects being taken upon the ground, it will plot the position of the observer without further measurement.

In construction the station pointer consists of three arms, generally about fifteen inches long, moving about one common centre. One of the arms, which may be considered as a base, is attached to a protractor. Each of the other moveable arms carries a vernier, which reads upon the protractor, and may be set to any reading required. The centre is perforated with a small hole, that will admit a pricker through it, to mark its position on the drawing.

To lay down a Position with the Station Pointer.— The arms are set to the angles subtended by two ob-

jects already ascertained, in relation to a third object
or base. The instrument is then placed upon the plot,
so that the arms bisect the three known objects or
points. The centre will then indicate the position from
which the angles were taken, and may be pricked off
through the hole.

In marking from the divided edges of the kinds of
protractors that read down to the working surface described in this chapter, it is considered best to use a
needle-point for the purpose. The eye of the draughtsman should be placed at right angles to the lines on
the surface of the protractor. The extension of the line
from the protractor to the needle-point should be
sighted from the centre of the protractor. When the
needle is removed, after the mark is made, the hole
should be again observed, to see if it coincides with
the direction of the line. If it does not do so, the
pencil line may be subtended from the centre a little
on one side or the other of it.

CHAPTER XXXI

INSTRUMENTS FOR COMPUTING THE AREA OF SURFACES
OF DRAWINGS—COMPUTING SCALE—PLANIMETER—
OPISOMETER.

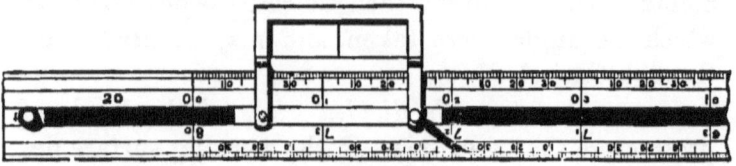

Computing Scale.

The COMPUTING SCALE, or *Computer*, as it is called, is at the present time almost universally used for computing the area of land from plans. It is as ingenious as it is simple, entirely superseding the laborious trigonometry of the past, also effecting a saving of two-thirds of the time required by any other method.

The computer consists of a scale of boxwood, generally twenty inches in length; along the centre is an undercut groove, in which a short slip of metal is fitted to slide loosely. A light metal frame, placed by the side of the scale, is attached to the sliding-slip by two light blades of metal. A small handle, fixed over the slip, moves the frame to any part of the divided scale. Across the centre of the metal frame is a line, which is the index to read off the quantities upon the plan.

The index line in some computers is drawn upon a piece of horn fitted in the frame; the horn becomes cloudy and cockled, which renders the plan beneath it very obscure. The author employed glass for the interior of the frame; this is also in some particulars objectionable. A gentleman connected with the Tithe Commission Office, where perhaps the greatest amount of computing is performed, has lately suggested to the writer the employment of a fine needle for the index line; this appears to answer better than the many previous experiments.

The divisions for reading off the computer are generally placed along the centre of the scale, by the side of the groove, the acres and roods being divided upon the scale, and the perches upon the sliding-slip. In the illustration, which represents a part of the author's improved computer, the scales are fully divided to the edge, and are read off by one observation, to a line upon the edge of the frame.

In some computers the scales are figured to read both to left and right hands; they are much better if made to read from left to right only, the slide to be stopped when it arrives at five acres. When there are two scales on the computer, as two and three chains, which is commonly the case, only one scale can stop at the five acres. To obviate this, a loose piece of metal, to form a stop to the second scale, will be found convenient.

In computing a large piece of land, every time the index arrives at five acres, it is customary to make a mark upon a piece of paper as a memorandum. The author has sometimes put a small ivory *teller* to indicate how many times the index has been to the five, which saves this trouble; but as it adds to the expense of the instrument, it will be considered by many an unnecessary refinement. It may be useful to persons who are not frequently computing, and who may lose much time by possibly forgetting to make a mark for one of the *fives* when absorbed in the operations of the instrument.

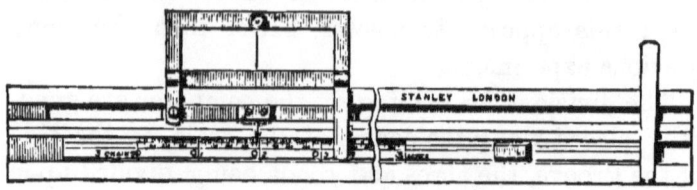

Universal Computing Scale.

The writer has recently made for H.M. Tithe Commission Office a *universal* computing scale, the design of some of the gentlemen in the office. This computer is one in which the division is placed on a separate slip of boxwood. The slip slides into an undercut groove, and is read by an index. By this plan one computing frame only is used for as many scales as may be required.

Before commencing to use the computer, it is necessary to be provided with a sheet of transparent material, ruled entirely over in one direction, with lines one chain apart, to the scale by which the plan was plotted. Tracing-paper or cloth may be used, but it does not answer very well for this purpose, being insufficiently solid. The material found practically best is what is termed *horn paper*. The preparation of this material is so little known that it can scarcely be purchased. The author's manner of preparing it for use with the computer is as follows: Obtain a sheet of stout paper of the kind used for type printing; it should be of rather loose texture. This should be damped and fixed upon a drawing-board. The distance of the lines, one chain apart, should be carefully pricked off along one edge of the paper, and lines ruled

entirely over it from the punctures by a tee-square or parallel rule. To render this transparent, it is necessary to pin it to an open frame and varnish it thoroughly from both sides several times with very good copal varnish. It will require to be kept a month or longer before it is fit to use. If properly prepared, one piece will last many years in constant use. Some horn paper that the author has seen in the Tithe Commission Office has been in constant use eight or nine years. This appears of better quality than he has been able to prepare by the above method.

To compute areas with the computer, the horn paper is first placed over the plot of which the area is required, and secured by leaden weights, that it may not slip. By this operation the plot of land, as seen through the horn paper, appears to be divided into slips of one chain wide. If we now measure off the united lengths of the whole of these slips within the plot, we readily obtain the contents. This is done by the computing scale, as follows: The computer, with its index at zero, is placed parallel with the lines upon the horn paper, in such position that the index will commence reading from the left-hand end of the first slip of land, as it appears through the horn paper; it is held in this position while moving the index along the scale until it reaches the right-hand end of this slip; thus it will take off the length of the slip upon the computer. The instrument is then lifted, being careful not to shift the index, and the index is placed over the left-hand end of the next slip of land; then, by moving the slide along the scale to the right-hand end of this slip, the second slip will be added to the first, and so on, until the measurement is completed.

If the piece contain more than, say, five acres, which will be the end of the computing scale of 2 or 3 chains to the inch, a faint mark may be made with a soft pencil upon the horn paper, unless some object appear at the spot on the plan where the computer stops ; a mark should also be made upon a piece of waste paper to indicate that five acres have been computed. The index may now be brought back to zero upon the scale, and the computing continued from the mark on the horn paper until the index arrives at another five acres, or at the completion of the plot.

In the description just given, the index is said to be placed over the commencement of the slip of land, as it appears through the horn paper, which can only be done if the plot is right-angled. If the sides of the land are oblique, the needle forming the index is so placed upon the boundary that there appears an equal quantity of land excluded to that taken in the measuring. The same observation will apply in placing the horn paper over the plot. Many persons compute plots with the horn paper placed diagonally, which is perhaps the best way if the figure is irregular, or it will not read easily into equal chains one way. The reading of the scale of the computer shows in figures the actual quantity in acres, roods, and perches that are contained in the slips of land the index has passed over.

It may be thought that there would be some difficulty in equalizing a space within and without the line, but practice tells us there is no difficulty to moderately intelligent observation. In the early use of the computer it was tested against the best trigonometrical measurements of plans that could be found; and I have the authority of Colonel Leach, the well-known director of the Tithe and Enclosure Commission Office, that in all cases its results were superior to the trigonometrical measurements.

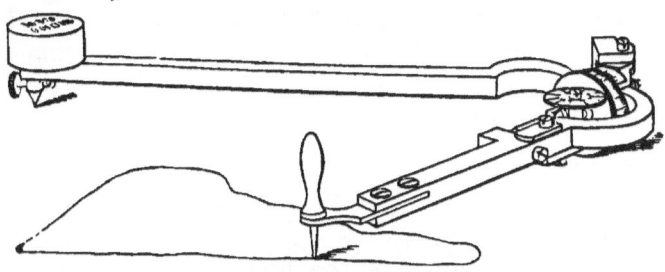

Amsler's Fixed Scale Planimeter.

THE POLAR PLANIMETER is a very exact scientific little instrument for measuring areas, either by actual measurements or to proportional scales. It was invented by Jacob Amsler, a Swiss professor of mathe-

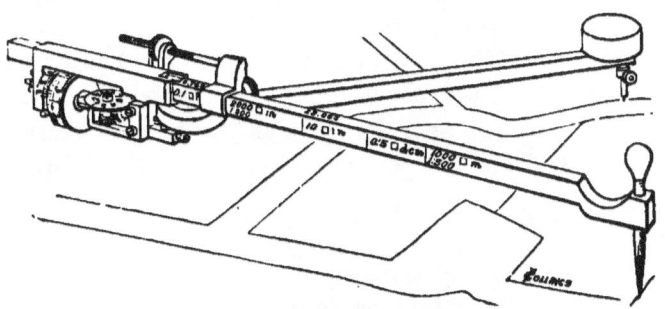

Proportional Polar Planimeter.

matics. It is altogether the best and most exact instrument for the computation of such areas, as for railway sections, steam pressure diagrams, and other small work. It is also the most exact instrument for computation of very irregular outlines which cannot be reduced to lines or approximately regular curves, as indented and irregular coast lines to small scales. The instrument is perfect, subject to the possibility of the hand following exactly the outline of the figure to be computed; but practically is less expeditious, and not more exact for computing large surveys of land, to moderately large scales as three chains to the inch, as it proved under trial by the talented computers attached to the Tithe and Enclosure Commissions some years ago.

Planimeters are made of several forms, the two kinds illustrated upon the last page being the most general.

The *Fixed* Planimeter, the first illustration, represents the instrument as made to one scale only, for this country, in square inches of actual measurement. It can be applied to steam indicator diagrams directly, or to computation of land, by a multiplier of the square of the number of units to the inch, as 4 for 2-chain scale, 9 for 3-chain, 16 for 4-chain, etc.

The *Proportional* Planimeter is shown in the second illustration. In this the unit can be changed by altering the radius of the arm that carries the tracer to any of the scales divided upon the planimeter, which are as follows:—

1 sq. dcm.	= 6, one square decimetre	
0·1 sq. f.	= 0·1 square foot	
2000 sq. m. }	= 2000 square metres	Every total
1 : 500	on a scale 1 : 500	rotation of
10 sq. in.	= 10 square inches	the roller.
0·5 sq. dcm.	= 0·5 square decimetre	
1000 sq. m. }	= 1000 square metres	
1 : 500	scale 1 : 500	

The details of construction are,—

A bar, carrying a fine needle at one end, that is made to enter the surface of the plan on which the plot to be measured is drawn; this bar forms the point of attachment of the instrument, and may be of any convenient length. At the opposite end to the needle, which is cranked to escape contact with some of the apparatus when in movement, a pair of vertical pivots are centred, so that the bar can move freely upon them in a horizontal direction only. The axes of the pivots are carried either upon the principal part of the instrument, *the tracing arm*, as in the first engraving, or upon a moveable sliding fitting upon it, as in the second illustration. The tracer, at the end of this arm, is used to follow the outline of the figure to be measured. Upon the tracing arm or, if sliding, on the fitting, the measuring apparatus is fixed. This consists of a roller on a pivoted axis, which must roll very freely. The roller carries a drum, divided into 100 parts, reading into a vernier, which gives the reading of the drum's revolution to the $\frac{1}{1000}$ part of its circumference. Upon the same axis as the roller an endless screw is cut, that works into a worm wheel of ten teeth which records the revolutions of the roller. The proportional scales of the instrument are derived from the distance of the vertical axis of the tracing arm multiplied into the circumference of the roller. Upon the top of the tracing arm a series of constant numbers are engraved, which vary slightly with the construction of the instrument; in the one before the writer they are 20·781, 20·769, and 22·065, which are units of complete circumscribed areas when the tracing point is within the figure to be measured.

To use the planimeter. 1. If of the *fixed* kind, the reading will be in inches. If of the *proportional* kind, this has first to be set to the scale. For this last, one of the divisions upon the bar is taken according to the nature of the computation; thus, if the area is required in inches, 10 ☐-in. will be the most convenient. To this the line on the bevelled edge of the slide upon the bar is set, and the instrument is ready for use. There is a clamp and fine adjustment to get this to position exactly. It may be observed that the quantity produced by working the instrument will be given in inches. If the scale is to fractional parts of an inch it is multiplied by this fraction decimally from the units contained in the reading as before mentioned.

2. Place the instrument upon the paper so that the tracing point, roller, and needle point all touch the surface at any convenient position. Press the needle point down gently, so that it just enters the paper, and place the small weight over it.

3. Set the tracing point to any part of the outline of the figure to be computed, and make a mark. Before commencing, read off the counting wheel and the index roller. Suppose the counting wheel marks 2, the roller index 91, and the vernier 5, write this down 2·915.

4. Follow with the tracing point exactly the outline of the figure to be measured in the direction of the movement of the hands of a watch, until you arrive at the starting point; now read the instrument. If there are straight lines in the figure, these may be traced along a slip of horn. We will suppose this reading to be 4·767. We have now completed the area as far as the instrument is concerned, but there are now three points to be considered.

Firstly. If the needle point was outside the figure when it was traced, we deduct the first reading, 2·915, from the second, 4·767; the remainder, 1·852, indicates that the measured area contains 1·852 units, the value of the units depending on the setting of the bar, which was 10 □-in. We have thus 1·852 × 10 □-in. = 18·52 inches—the exact area of the figure measured.

Secondly. If the needle point was inside the figure when it was traced, which it is necessary to be if the figure be large, then the number engraved at the setting on the *top* of the bar is added to the *second* reading and the first reading is subtracted from it. Thus, suppose the readings to be as before:—

Second reading	4·767
Number engraved above 10 □-in.	22·141
	26·908
Deduct first reading	2·915
	23·993
Multiply by 10 □-in.	10
	239·93 inches.

Thirdly. The counting wheel may have gone through more than one revolution forwards or backwards. If forwards, as 9, 0, 1, 2, etc., then, as many times as the zero passes the index line, add 10,000 to the second reading; but if moving backwards, as 2, 1, 0, 9, etc., then add 10,000 to the first reading.

Mr. F. J. Bramwell, C.E., has given a most excellent paper on the polar planimeter, which is printed in full in the "British Association Reports" for 1872, page 401. Until the appearance of this paper, no one appears to have been able to describe its principles of action so intelligibly as to be comprehensible. As it is given in this paper of about eleven pages, with eleven engravings, the matter is most clear, and to those interested the writer would strongly recommend reference to it. In former editions of this work, the subject appeared quite formidable, as only possible of description with much more space than could be devoted to it. With the knowledge of Mr. Bramwell's schemes for demonstration it now appears much less so, therefore the following description will follow Mr. Bramwell's plan, as nearly as the uniformly condensed descriptions of this work permit.

First, as to the *Roller*:—This is centred with perfect accuracy with its axis in a line parallel to the arm. Therefore if the arm were dragged along the surface of paper with the roller touching in its own line, that is, parallel to the axis, it is quite clear that the roller could not revolve. It is also clear that if the arm were moved transverse to itself, that is to the axis of the roller, it would be in a position similar to the axle of a carriage, and the roller would roll, and the circumference of it would measure the distance that the axle moved. Now this line of motion which is transverse, and line of no motion which is longitudinal, entails, that if the roller were moved obliquely, it would roll the amount of transverse motion that there might be in the *obliquity*, and slip the amount of longitudinal motion that it would contain. Therefore, supposing the lines A B C and A' B' C' is the axis of any roller whatever in two positions, if this axis moves downwards from A to A' such a part of the circumference of the roller will move on the surface as is contained

in the line A to A'. If the axis is moved so that the
roller travels obliquely from B to B', as we attain no

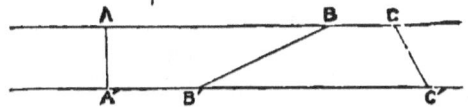

movement of the roller by longitudinal displacement,
this would register exactly the same as A to A', that
is, the amount of transverse movement only; or C to
C', would be the same. That is, supposing the axis
kept always parallel to the lines A B C and A' B' C'.
Therefore the lines A A', B B', and C C', are equal, as
far as the revolution of the roller is concerned, and the
motion of this is what we extract the area from.

We will now construct a very elementary planimeter,
which for our purpose may diagrammatically be represented by a single line of given length, and we will
measure a figure which shall contain ten square inches.
Our planimeter shall be of exactly five inches in length,
and a roller shall be placed upon it in the axis of
exactly two inches diameter. By this, if our *line
planimeter* passes with its roller over the distance of
the entire circumference of the roller, the area of the
parallel space between its first and second positions will
contain exactly the ten inches we desire to measure,
that is, five inches by two inches.

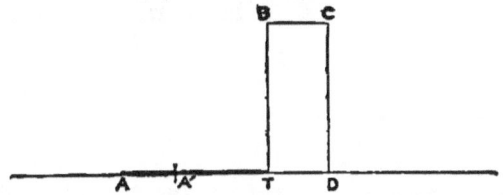

In the above diagram let A to T represent our
planimeter, and be five inches in length, T being our

tracer.   Let A′ represent a roller of two inches
diameter, and T B C D a parallelogram to be measured.   We will call A to D our base line.   Now, if we
move the tracer T up the line to B, keeping A always
constant on our base, our planimeter will stand vertical
to the base when the tracer reaches B.   If we now
look at the roller, it will have registered a certain quantity which is of no consequence to us, for reasons to be
given.   We now move the tracer B to C, by which,
as it was only of the length B to T, that is, equal to
A to T, for one point to keep on the base it must
move parallel, and our planimeter will therefore stand
again vertical at C.   Therefore, in the space B to C
the roller will make one revolution, or two inches.
This last is all the dimension we really require of it,
as our planimeter was five inches, and the roller two
inches, the inscribed figures $5 \times 2 = 10$ give the
quantity required, which we could have represented
by the division of the roller into 10 equal parts at
once.   But we have not yet completed our figure by
moving entirely round its circumference, and we have
neglected the quantity moved T to B; we must therefore watch the return journey of the tracer back to T,
its starting point.   If the tracer follow down the line
C to D, keeping the other point of our planimeter constantly upon the base, we may observe that this last
motion of the instrument is exactly the reverse of the
motion on the first line, T to B, which was upwards.
Therefore, as regards the motion of the roller, whatever quantity it moved in going T to B, it must reverse
this in going C to D; therefore, as regards the tracing
of these two lines, in this instance they produce no
effect on the reading of the instrument, for whatever

one winds on the roller the other unwinds. We
have now only to move the tracer back D to T, which
being exactly longitudinal will cause no motion of the
roller. Therefore by the entire circumference of the
figure we have simply measured the space B to C,
which, multiplied into the radius, gives us ten inches,
the quantity shown on the roller by one revolution.
We may observe in this, that if we had not carried our
tracer so high as C, it would not have been transverse,
and the area would have shown that on the roller also, if
we had returned on the line C before we had reached
D, the area would have been less.

We may now take the same diagram and lay the
parallelogram flatways,—

Take as before for our planimeter the line A to T.
The lines T to C and T′ to D being equal and reverse,
may be neglected as before, as also the line of no

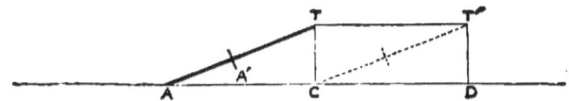

motion, C to D. The amount of motion in the roller
travelling from T to T′ will therefore be only the
amount of transverse motion that is given to the roller,
which is equal to T to C, upon the proposition given
at page 252, therefore in moving this distance the roller
will make one revolution of two inches, as before;
which, multiplied into the radius of the planimeter,
would give ten inches, as before.

It would appear from the two schemes above, that
parallelograms of any figure would certainly follow the
rule. Now, for the construction of any other form we
can imagine any area to be composed of an infinite

number of parallelograms to make its outline represent any figure whatever, therefore the instrument is true for all forms.

In the above schemes a straight line is taken for the base. In the instrument it is an arc, derived from the constant radius of the arm by which the instrument is attached to the paper. The length of the arm performs no function to the measurement, and might be extended to indefinite lengths, so that its circumference might be representable by the straight line base we have considered; the radius of the tracing arm being to the tangent of the circle constant, it is as the base line considered.

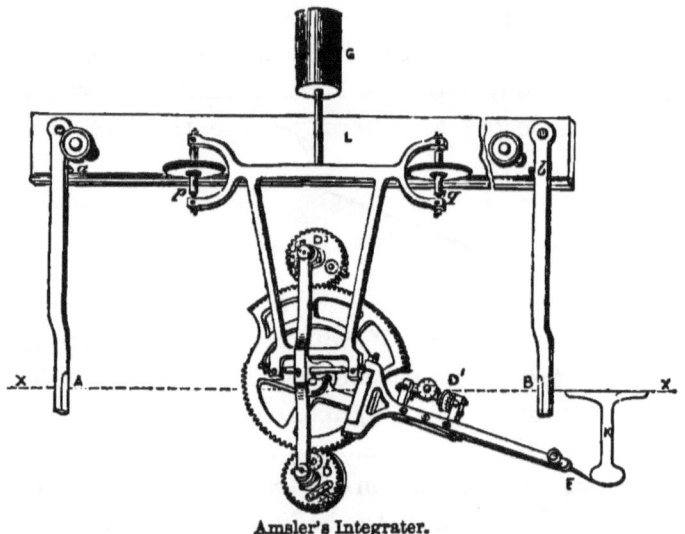

Amsler's Integrater.

The INTEGRATER is another invention of Mr. Jacob Amsler, who has given me the following short description. This instrument, besides computing areas of greater magnitude than the polar planimeter, from its greater mobility gives also the static momentum and inertia of a plane surface taken in relation to any axis whatever.

The apparatus consists of a railway, upon which the instrument proper is placed, which it directs in one parallel, and in the plane of the figure to be calculated. A carriage is placed upon the railway, the wheels of which move in the grooves of the rails. To the carriage are jointed several wheels upon the plane of the carriage, which are best shown by the engraving. One arm carries the tracer F.

In tracing the outline of a figure by the tracer, the carriage makes a movement forwards and backwards during the time that the wheels make a movement combined of sliding and rotation as with the planimeter. The final rotation to the wheels may be observed by the divisions engraved upon the drums and the calculating system immediately connected with it, as in the planimeter.

If u be the rotation of the wheel system D^1,

,, v ,,   ,,   ,, D^2,

,, w ,,   ,,   ,, D^3,

we shall have the contents of surface $= u$,

The static momentum $= a\,v$,

The moment of inertia $= u - b\,w$,

a and b representing the constants, which depend upon the dimensions of the apparatus. For the diameter, taken as unity for the instrument illustrated, we have,

$$a = 0\text{·}6. \qquad b = 0\text{·}4.$$

The axis to which these moments are relative is a right line passing through the centre parallel to the directing rails, upon which the carriage moves.

· The integrater possesses the qualities, that it is only necessary to follow once the contour of the figure, to

s

find the three quantities sought, by the separate read-ings.

If the movement of either of the calculating systems is retrograde, it must be read as a negative number, as with the planimeter.

OPISOMETER.—For roughly measuring distances upon maps, the instrument illustrated below may be used. It consists of a small milled-edged wheel, which re-volves upon a screw for an axis. The screw moves through the guides between which the wheel is placed, being propelled by the revolution of the wheel, until the wheel has drawn the screw to one of the sides, and is stopped by a collar at the end of the screw.

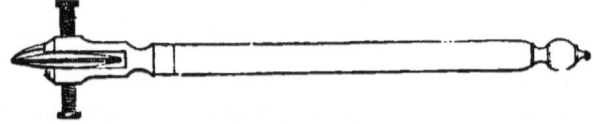

Bennett's Improved Opisometer.

To prevent the screw turning· with the action of the wheel, a groove is made longitudinally up it, in which a pin projecting from the bearing is fitted.

To measure a line, either straight or irregular, on a map with the opisometer, the wheel is first turned until it is stopped by the screw bringing the collar against the side ; the wheel is then run slowly along the line the distance required to be measured. This distance may be afterwards ascertained by running the wheel the reverse way along a scale of quantities from 0, until it is stopped by the collar being brought back as at starting—the position at which it stops upon the scale indicating the length of the line passed over on the map.

The above opisometer is sometimes constructed with

the screw to run on a scale, so that measurements may be taken with it direct.

There is another construction of opisometer, which is cheaper but not quite so good as that described above.

CHARTOMETER AND WEALEMEFNA.— These two very useful little measuring instruments, lately invented by Mr. E. R. Morris, have become very popular. The chartometer measures approximately any quantity to scale by rolling the small milled wheel, at the bottom of the instrument, over the drawing or map. All the ordinary scales for plans of lands are supplied with the instrument. The scales, which are circular card disks,

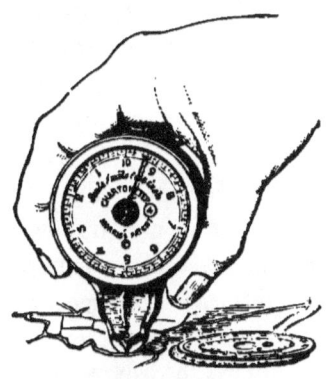

Chartometer.

are made to exchange, as required, by opening the face of the instrument.

The wealemefna is simply a measurer to feet and inches. It is quite a curiosity in its way, being only about the size of a shilling, and measuring off 25 feet by rolling contact of the little roller, with very passable accuracy.

CHAPTER XXXII.

DRAWING PAPER AND METHODS OF FIXING IT—TRACING
PAPER AND CLOTH—CARBONIC AND BLACKLEAD PAPER
—DRAWING PINS—PIN LIFTER—STATIONER'S RULE
—CUTTING GAUGE—LEAD WEIGHTS—VARNISHING,
ETC.

IN this and the following chapters, which may be considered as an Appendix to the work, the articles to be described partake more of the nature of drawing materials and utensils than of mathematical instruments. It is presumed, however, that the usefulness of the subject will be a sufficient apology for introducing it.

DRAWING PAPER.—Of this we can say very little, only that it is important to the draughtsman to have a suitable surface to work upon. The only good drawing papers that have come to the notice of the writer are those known as Whatman's. These are so generally used as to need no comment. There are two distinct kinds of drawing paper in use, one called, technically, *not*, and the other *rough*. The *not* paper is best suited for mechanical or elaborate architectural drawings; the *rough* is more effective for architecture, perspective, or Gothic elevations. This, however, is a matter of taste, as either kind works equally well. There is yet another description of drawing paper, termed *hot-pressed*. This is seldom used, as it does not take colour so well as the *not* or the *rough*.

The size, and sometimes the make, of each paper has a distinct name. The papers considered best, and almost universally used, are the Antiquarian, Double Elephant, and Imperial; if smaller sizes are required, the half or quarter sheet is used. The larger size paper, *Emperor*, is also occasionally used for large plans or competition drawings. In practice it will be found most convenient to adhere to these sizes, as drawing-boards and tee-squares may always be had to correspond with them. The following table contains the dimensions of every description of drawing paper.

Demy	. . .	20 inches by $15\frac{1}{4}$ inches.	
Medium	. . .	$22\frac{3}{4}$,,	,, 17 ,,
Royal	. . .	24 ,,	,, $19\frac{1}{4}$,,
Imperial	. . .	30 ,,	,, 22 ,,
Elephant	. . .	28 ,,	,, 23 ,,
Columbia	. . .	35 ,,	,, $23\frac{1}{2}$,,
Atlas	. . .	34 ,,	,, 26 ,,
Double Elephant	.	40 ,,	,, 27 ,,
Antiquarian	. .	53 ,,	,, 31 ,,
Emperor	. . .	68 ,,	,, 48 ,,

For making detail drawings, an inferior paper is generally used, termed Cartridge; this answers for line drawings, but it will not take colours or tints perfectly. Continuous cartridge paper is also much used for full-sized mechanical details, and some other purposes. It is made uniformly 53 inches wide, and may be had of any length by the yard, up to 300 yards. It is, surface and strength considered, one of the cheapest papers made. There is also a kind of surfaced calico, which can be drawn upon very nicely for details, called *indestructible cloth*. It is about as cheap as good drawing paper, and can be recommended

for work to which constant reference is required, or
for permanent mechanical engineering manufactures,
which are repeated at periods.   It is also very serviceable and portable for diagrams.

Cottam's water-colour drawing tablets are sheets of
Whatman's paper laid on ordinary pasteboard mounts.
These are very nice for architectural and engineering
works, they lie quite flat in a drawer, and are ready to
be shown any time.

For plans of considerable size, *mounted paper* is
used; or the drawings are afterwards occasionally
mounted on canvas or linen.   The mounting of paper
is a business requiring considerable skill, and is done
at so moderate a price, that the amateur will save
nothing by doing it himself.   For mounting drawings
on paper of ordinary size, the process is simple.   The
linen or calico is first stretched by tacking it tightly
on a frame or board; it is then thoroughly coated
with strong size, and left until dry.   The sheet of paper
to be mounted requires to be well covered with paste;
this will be best if done twice, leaving the first coat about
ten minutes to soak into the paper.   If two sheets are
done at once, it is better that they be laid the pasted
sides together to soak, which equalizes the pasting.
After the application of the second coat, it must be immediately placed on the linen, and be dabbed all over
with a clean cloth.   It must not be drawn or stretched,
and it should not be cut off until it is thoroughly dry.

TRACING-PAPER AND CLOTH.—In practice, finished
drawings are seldom worked from, as they would
speedily become dirty and obliterated, and dimensions
would be taken from them with difficulty.   To obviate
this, copies of drawings are generally made on tracing-

paper or cloth, which are transparent materials too well known to need particular description. Either of these materials may be drawn upon by following the outlines of the original drawing, placed under it, most conveniently by employment of a rolling parallel rule. Tracing cloth is to be recommended for durability; it is sold in continuous lengths of twenty-one yards, and may be had from eighteen to forty-one inches in width, so that it is adapted to the drawing papers mostly used. That known as *Sager's vellum cloth* is of very excellent quality, both for transparency and strength. There is also a very stout tracing cloth called *indestructible,* very good for details. Tracing-paper varies considerably in quality. That is best which is toughest, most transparent, and freeest from greasiness. The continuous papers are more economical than those in sheets, as just the quantity required can always be taken from the roll.

Drawings of an ornamental description, illuminated, etc., are sometimes reproduced, to obtain a finished copy from a first draught which has become soiled and marked in designing. This is accomplished by what is termed a *transfer*. The common method is to fix a sheet of blacklead paper, with the blackened side downwards, between the original and intended copy, and to pass over the outlines or angles with a tracer, as described at page 19, and to reproduce the finished drawing from these lines or points.

BLACKLEAD PAPER is prepared by rubbing thin paper over with a soft block of Cumberland lead. It may be purchased properly prepared on suitable paper in single sheets at a trifling cost.

CARBONIC PAPER is a blue paper which has one side

painted over with lamp-black, ground to perfect fineness in slowly drying oil, and left a considerable time
to season. It is used in a similar manner to blacklead
paper, but for coarser purposes.

SECTIONAL PAPER, for sketching to scale or for making small working drawings, is very convenient. This is
paper ruled over into small squares to a given scale
with pale ink. The spaces in ordinary use are $\frac{1}{10}$,
$\frac{1}{8}$, $\frac{1}{6}$, $\frac{1}{5}$, and $\frac{1}{4}$ inch. Thicker lines are put either to
mark off the inches, or to count the spaces in tens.
In using this paper the scale is dispensed with, the eye
being quite sufficient to subdivide the spaces when
part only of the quantity represented by one space is
required. This paper is also made up into sketching-
books and architects' pocket-books; for the latter it is
particularly convenient.

Drawings have frequently to be made upon parchment for plans upon deeds, specifications for letters
patent, etc. A special parchment can be had for this
purpose. There is also a kind of parchment made
that is quite transparent, which can be purchased cut
to the Patent Office regulation size. Before inking or
colouring upon parchment, it should be pounced over
with pouncet of finely powdered French chalk. It will
even then sometimes work greasy; if it should do so, a
little ox-gall in the ink or colour will remedy it.

There are several methods of attaching drawing paper
to the board; one of the most simple is by means of
drawing-pins, which are merely pressed through the
paper into the board.

The DRAWING PIN, as shown next page, is a small flat
piece of turned metal, with a moderately stout needle-
point screwed into the centre of one side. The upper

side of the head of the pin should be a portion of a sphere, the edges being thin, so as to lie close to the

paper, otherwise it will injure the tee-square when passing over it. The steel point of the pin should not be too much tapered, or it will be continually flying out. For securing tracing-paper, pins with large heads should be used.

The PIN LIFTER will be sufficiently explained by the following engraving; it saves finger-nails and penknives. It is merely a small chisel bent to act as a lever when

Pin Lifter.

the edge is inserted under the head of the pin, to draw it out.

 The French have recently introduced an angular pin, which has three points on the under side, as shown in the illustration; it will secure the corner of the paper very firmly.

If the paper is to receive an elaborate drawing with colour, etc., it is necessary to attach it to the board with some kind of cement. Glue is undoubtedly the best. The following is the usual manner of proceeding. The stretched irregular edges of the sheet of paper are cut off against a stationer's rule, squaring it at the same time. The sheet of paper is laid upon the board the *reverse* side upwards to that upon which the drawing is to be made.   It is then damped equally over, first by passing a moist clean sponge, or wide brush, round the edges of the paper about an inch and a half on, and afterwards thoroughly damping the whole surface except the edges.   Other plans of damping answer equally well; it is only necessary to observe that the edges of the paper should not be quite so damp as the other parts of the surface. After the paper is thoroughly damped, it is left until the wet gloss entirely disappears; it is then turned over and put in its position on the board.   About half an inch of the edge of the paper is then turned up against a flat ruler, and a glue-brush with hot glue passed between the turned-up edge and the board; the ruler is then drawn over the glued edge and pressed along.   If on removing the ruler the paper is found not to be thoroughly close, a paper-knife or similar article passed over it will secure perfect contact.   The next *adjoining edge* must be treated in like manner, and so on each consecutive edge, until all be secured.   The contraction of the paper in drying should leave the surface quite flat and solid.

Some draughtsmen dip a piece of glue in hot water and rub it along the turned-up edges which are to be attached; this is a somewhat tedious and uncertain method.   It is better to have a small glue-pot, which should have a cover to keep the glue clean and moist while heating.   It is necessary that the glue be kept quite thin by adding water to it every time it is heated, as it rapidly thickens by evaporation.   To replenish the glue-pot, the cake-glue should be first soaked in cold water for at least eight hours.   The better the quality of glue, the more water it will absorb, and the thinner it may be used, which is an advantage, as thick glue will be found to get under the paper farther than is required, by the action of rubbing the edge down.

To remove the drawing from the board after it is finished.—A very ingenious and expeditious method has lately been introduced to the notice of the writer, which is by means of a cutting gauge, of which an engraving is given below. It is suitable for all full-size drawings, and does not injure the board in any way, as the cutting-point is only allowed to sink the depth of the thickness of the paper to be cut. The gauge is similar to those used by joiners, and consists of a stock of wood having a mortise in about the centre,

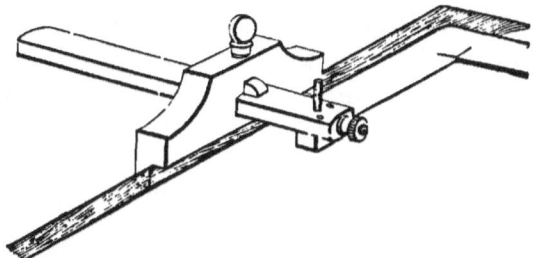

Cutting Gauge to remove Paper from Drawing-board.

through which a rule carrying a cutting-point passes to the required distance, where it may be clamped by a screw in the stock. The manner of using the gauge is to set the cutting-point at the distance off the stock that the cut is required from the edge of the drawing-board, then to draw the stock along with pressure against the edge of the board, at the same time throwing sufficient weight upon the cutting-point to separate the paper. The gauge will only cut conveniently within a short distance of the edge of the board.

The general method in practice, to cut off drawings,

is to make a pencil line round the paper with the tee-square at a sufficient distance to clear the glued edge, and to cut the paper with a penknife, guided by a stout ruler. In no instance should the edge of the tee-square be used to cut by. A piece of hard wood, half an inch thick by two inches wide, and about the length of the paper, forms a useful rule for the purpose, and may be had at small cost. The instrument used for cutting off, in any important draughtsman's office, is what is termed a *stationer's rule,* which is a piece of hard wood of similar dimensions to that just described, but with the edges covered with brass. It is neces-sary to have the edge thick, to prevent the point of the knife slipping over. Either of the above rules will also answer to turn the edge of the paper up against when glueing it to the board.

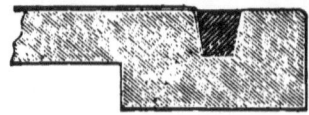

Mr. J. Stoney, C.E., has devised a means of holding the paper stretched upon the board, which is simple, and answers very well. The board is constructed with a taper groove, as shown in section in the illustration above, entirely round the board. Into this groove four lengths of wedge-section slips are fitted, one for each edge. In use, the paper is first damped, then turned down into the grooves, and the wedge-slip is inserted upon the paper, this draws it down and holds it tightly. The wedge-slip is rendered nearly tight by the pressure of the hand, but a few taps of a mallet will render it perfectly so. To remove the wedge, after

the paper is cut off, there are holes through the board, along under the wedge-slip, in which a wooden punch can be inserted, and a slight tap will set them free.

When tracings are to be made from plans, drawings, or portions of drawings, it is often found convenient to hold the tracing-paper with lead weights. These often enable a small piece of tracing-paper to be used when only a small portion of the drawing is required to be reproduced; and this without injury, as would be the case if the drawing pins were used. Lead weights are also convenient for holding open for examination

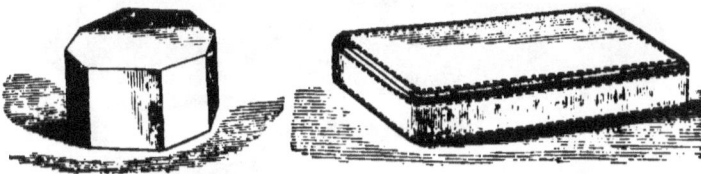

S. & A. Dept.   Lead Weights.   T. C. Office.

drawings that have been rolled up. A very neat kind of weight, made of mahogany, of octagon form, is used in the Science and Art Department at South Kensington; in this the inside is turned out and loaded with lead. Another excellent weight, of oblong form, neatly covered with solid calf leather, is used in the Tithe Commission Office. Lead weights in any instance require covering, otherwise they mark the paper and soil the fingers.

Large plans of estates and other large drawings are occasionally required to be varnished. This is better if done by a practised hand, or by the mounter, as it requires both care and skill. The process generally followed is to stretch the drawing upon a frame, and to give it three or four coats of isinglass size with a flat broad brush, observing to well cover it each time, and to allow it to dry between each coat. The varnish to be applied is composed of Canada balsam diluted in oil of turpentine. This requires to be put on evenly in a flowing coat, and left in a warm room free from dust until thoroughly dry.

Complete sets of drawings of meritorious works are valuable in themselves, both as examples and me- mentoes. They may be kept very conveniently in long cylindrical, japanned tin cases, which are not expen- sive, and are perfectly dust-tight. The cases, if neatly written upon and placed in a rack, form a businesslike ornament for the draughtsman's office.

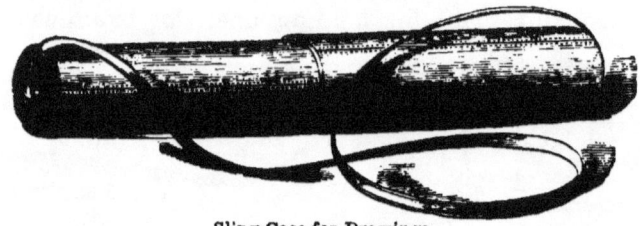

Sling Case for Drawings.

The most convenient method of carrying drawings for building works in execution, is to have a solid leather case similar to a telescope case. This is best if made with the cap or lid of the same length as the body; it can then be drawn out any distance according to the length of the rolled drawing. If thought more convenient, and the drawings are heavy, a strap may be added to pass over the shoulder.

Portfolios, when used for drawings of large size, as antiquarian or double elephant, should have the holland flaps to close together with an elastic strap, to support the covers. A pair of neat oak battens screwed outside each cover does not look unsightly, and will keep the portfolio in good condition.

CHAPTER XXXIII.

It is well known that no drawing-pencil is quite so good as one manufactured from plumbago sawn from a fine block of Cumberland lead; this is firm, black, and brilliant, and will erase perfectly; the last quality is possessed by no other pencil. The difficulty is to get pencils made of this fine quality; the greater part of the Cumberland-lead pencils which are made from the native lead, are unequal in hardness and colour, and frequently gritty. Very many draughtsmen prefer to endure these occasional faults for the sake of the excellent erasing qualities and the pleasant touch which they all possess, and which art has never produced in the most carefully prepared compositions.

The drawing-pencils mostly used by professional men are of German manufacture; of these, Faber's hexagon pencils are perhaps the best. They are in five degrees of hardness. The good qualities of these pencils are perfect equality of firmness and colour; their only fault is the apparently greasy nature of the composition, which renders the marks difficult to erase.

Blacklead pencils may be had of several degrees of hardness, and of various qualities, which are generally expressed by certain letters of the alphabet in the following order, commencing with the hardest and finishing with the softest and blackest:—H H H H H H H H H, H H H, H H, H, F, H B, B, B B, B B B, B B B B, and

B B B B B B. The first two are used for drawing on wood only; the H H H and H H are used for plotting and mechanical drawing; the H H and H are used for architectural drawing; the F and H B are used for sketching, details, and writing; and the last five for heavy sketching and shading.

The writer has had some pencils made for him of what is termed washed plumbago,—that is, the particles, instead of being ground, are mixed with water, and those only sufficiently fine to remain suspended a considerable time are taken. The sediment is then mixed with thin mucilage, and when sufficiently dry consolidated by hydraulic pressure. These pencils are as fine as Faber's, and erase better. They are made in five degrees of hardness, and of oval section, which form has some advantages for line drawing. The pattern is shown in the engraving above.

Pencils are also made small to suit the sizes of the pipes of the various drawing instruments; they are of similar quality to the above; but the best article for the purpose is the plain lead of one of Faber's artists' pencils, described on the next page. The pipe of the instrument must be made to fit it.

There is some little art in cutting a pencil properly. To those unacquainted with the proper method, the following hints may be useful. If the point is intended for sketching, it is cut equally from all sides, to produce a perfectly acute cone. If this be used for line drawing, the tip will be easily broken, and wear thick; it is much better for line drawing to have a thin flat point. The manner of proceeding is, first, to cut the pencil, from two sides only, with a long slope, so as to produce a kind of chisel-end, and afterwards to cut the other sides away only sufficiently to be able to round the first edge a little; this will be better shown by the illustration. It is scarcely necessary to remark that a pencil cannot be cut properly without the knife is sharp. A point cut in the manner described, may be kept in good order for some time by pointing the lead upon a small piece of fine sandstone, fine glass paper, or a fine file; this will be less trouble than the continual application of the knife.

Some pencils, patented by A. W. Faber, termed *artists' pencils*, have the lead moveable, the cedar being merely a holder. The lead is nearly the length of the pencil, and is held firmly by a very ingenious clamping apparatus. They save the trouble of cutting away the cedar, and are also very economical, as the lead has only to be replaced to renew the pencil.

The following observations on the means of erasing lines may be useful to some draughtsmen. To erase Cumberland-lead pencil marks, native or bottle india-rubber answers perfectly. This, however will not entirely erase any kind of German or other manufac-tured lead pencil marks. What is found best for this purpose, is fine vulcanized india-rubber; this, be-sides being a more powerful eraser, has the quality of keeping clean, as it frets away with the friction of rubbing, and presents a continually renewed surface;

T

the worn-off particles produce a kind of dust, which is easily swept away. Vulcanized rubber is also extremely useful for cleaning off drawings. A preparation of vulcanized rubber and powdered glass, known as *Green's ink eraser*, is found useful to the draughtsman for erasing or lowering the tone of colour, or hard ink-lines, also for removing stains or dirty marks.

For erasing ink-lines, the point of a penknife or erasing-knife is commonly used. A better means generally, is to use a piece of Oakey's No. 1 glass-paper, folded several times, until it presents a round edge; this leaves the surface of the paper in much better order to draw upon than it is left from knife erasure. There is a fine kind of size, sold as Erasure Fluid, a little of which applied with a brush will be found convenient to prevent colouring running.

To produce finished drawings, it is necessary that no portion should be erased, otherwise the colour applied will be unequal in tone; thus, when highly-finished mechanical drawings are required, it is better to draw an original and to copy it, as mistakes are almost certain to occur in delineating any new machine. Where sufficient time cannot be given to draw and copy, the writer has found it a very good way to take the surface off the paper with glass-paper before commencing the drawing; the colour will then flow equally over any erasure it may be necessary to make.

Where ink-lines are a little over the intended mark, and it is difficult to erase them without disfiguring other portions of the drawing, a little Chinese white may be applied with a fine sable-brush; this will render a small defect much less perceptible than by erasure.

CHAPTER XXXIV.

INDIAN INK—COLOURS—BRUSHES—PALETTES—CHROMO-LITHOGRAPHS, ETC.

INDIAN INK, or, more properly speaking, Chinese ink, is used for producing the finished lines of all kinds of geometrical drawing. Being free from acid, it does not injure or corrode the steel points of the instruments. The genuine ink, as it is imported from China, varies considerably in quality; that which answers best for line drawing will wash up the least when other colours are passed over it. This quality is ascertained

in the trade, but not with perfect certainty, by breaking off a small portion. If it be of the right quality,

it will show, when broken, a very bright and almost
prismatic-coloured fracture.

There is some difficulty in obtaining very good Indian ink for geometrical drawings; the brand by which
it is often indicated cannot be depended upon always,
as the same brand is occasionally impressed upon the
soluble brown quality, suitable only for artists.   The
two brands above are generally of excellent quality in
Yutsung inks.   All ink should be used immediately
after it is mixed; if it be re-dissolved, it becomes
cloudy and irregular in tone, but with every care it will
still wash up more or less.   To avoid this defect, the
writer has dissolved fine Indian ink in a non-corrosive
chemical menstruum; this is sold in bottles in a fluid
state at moderate prices.   It is not however recommended, the carbon in all menstrua, has a property of
coagulating and becoming gritty.

The writer has lately produced some liquid drawing
ink which appears to answer well, but has not yet had
a long trial.   It is prepared entirely from vegetable
decoctions, has the property of remaining soluble, is
alcoloid and non-corrosible; is of a fine black colour,
and will not wash up, or re-dissolve, when it becomes
dry.   Upon the same principle other line colours are
prepared, as crimson for section lines, blue for
dimension lines, and a burnt sepia brown, which is
prepared by some architects for mediæval architecture.
These inks have also been made for mining engineers,
of twelve distinct colours, for marking transverse
systems of cutting.   The colours being all waterproof
when dry, plans may be used in the damp under-
ground.

Water Colours are used for colouring drawings,

the most soluble, brilliant, and transparent are the best; this particularly applies to those used upon plans and sections. The colour is not so much intended to represent that of the material to be used in the construction, as to clearly distinguish one material from another employed on the same work.

The following table shows the colours mostly employed by the profession :—

Carmine or Crimson Lake	For brickwork in plan or section to be executed.
Prussian Blue	Flint work, lead, or parts of brickwork to be removed by alterations.
Venetian Red	Brickwork in elevation.
Violet Carmine	Granite.
Raw Sienna	English timber (not oak).
Burnt Sienna	Oak, teak.
Indian Yellow	Fir timber.
Indian Red	Mahogany.
Sepia	Concrete works, stone.
Burnt Umber	Clay, earth.
Payne's Grey	Cast iron.
Payne's Grey and Indigo	Rough wrought iron.
Dark Cadmium or Orange	Gun metal.
Gamboge	Brass.
Indigo	Wrought iron (bright).
Indigo, with a little Lake	Steel (bright).
Hooker's Green	Meadow land.
Cobalt Blue	Sky effects.

And some few others occasionally for special purposes.

The writer has lately introduced a series of liquid colours, prepared to the required tints, for every description of professional colouring on plans, sections, and details. The discovery of the colours is the result of some hundreds of experiments, in which he has tried every known description of colouring matter obtainable by solution. These colours are perfectly fluid, alcoloid so as not to rust steel, and non-poisonous.

They mix with water freely, in any proportion, so that evaporation is easily made up.   They are contained in two-ounce bottles, in which the brush is dipped directly,

so that no palette is required.   The colour contained in each bottle is enamelled on the stopper, and the bottle is labelled with the material which the colour is intended to represent technically, not with the absolute colour, or colouring matter, as is usual.   The professional colours for civil and mechanical engineers and architects, or such as each profession requires, are the following :—Brick section; Brick elevation, red, buff, and yellow; Stone; Concrete; Earth; Slate; Oak; Deal; Mahogany; Walnut; Cast iron; Wrought iron; Steel; Gun metal; Brass; Copper; Lead; Zinc; Grass; Fallow; Wheat; Barley, etc.

These colours are also arranged in cases, as shown in the cut above, with twelve or sixteen colours suitable for one profession.   They are also made so as to colour freely on parchment for patent specifications and plans on deeds. The advantages anticipated for these colours are general. They save the loss of time and inconvenience of getting out and washing up palettes often, when only a small quantity of colour is required; they secure uniformity of technical tint, particularly for such colours as are generally compounded of two or more colours, as gun-metal, steel, etc.; they are very economical, not costing more at first than cake colours, and entailing absolutely no waste. Each colour gives one tint only; they are therefore not at all adapted to artistic drawings or perspectives.

In colouring plans of estates, the colours that appear natural are mostly adopted, which are produced by combining various cake colours. Elevations and perspective drawings are also represented in natural colours, the primitive colours being mixed and varied by the judgment of the draughtsman, who, to produce the best effects, must be in some degree an artist.

Perspective elevations are now frequently coloured in monochrome; for this burnt sepia is the most effective, to which for distance, tints, and skies, a little cobalt may be added, and for foregrounds a little Indian red; the last is effective, but is perhaps a little violation of pure *monochrome*.

Care should be taken in making an elaborate drawing which is to receive colour, that the hand should at no time rest upon the surface of the paper, as it is found to leave a greasiness difficult to remove. A piece of blotting paper placed under the hand, and, if the square is not very clean, under that also, will prevent this. Should the colours from any cause work greasily, a little prepared ox-gall, which is sold in small pots at a trifling cost, may be dissolved in the water with which the colours are mixed, and will cause them to work freely.

For taking sketches of buildings, landscapes, etc., which for various purposes are frequently required in colour, a small japanned tin sketching box of moist

Sketching Box of Colours.

colours will be found the most convenient means. The colours are dissolved by merely moistening them with a little water. When tints are required, sufficient colour may be conveyed with a moist brush to the inside of the lid of the box, which is enamelled to form a palette. There is a ring on the under side of the box for the thumb to pass through to hold it securely, and space inside the box just sufficient for the brushes: thus the whole is complete in itself and very portable.

With the moist colours it is convenient to have a japanned tin water-bottle and cups; the whole is made in a very cheap and portable form. The cups will hook on to the edge of the palette. In use they are both partially filled with water, and used consecutively: one to wash the brush, and the other to dip for colour.

BRUSHES.—For colouring line drawings the best kinds
are made of sable hair.   There is a difference of opinion
among draughtsmen whether the light, termed red
sable, or the dark, termed brown sable, is the best.
The most important consideration is, that the hair
should be of good quality, which is judged of by
its length,—the longest and stiffest being the best;
nevertheless, the greater portion of the hair should
be in the quill.   It is only from being able to see
the length of the hair that the sable brushes are considered better in quills than in metal ferrules.   In all
instances a good brush should, when thoroughly
moistened with water, form a well-defined conical
point.

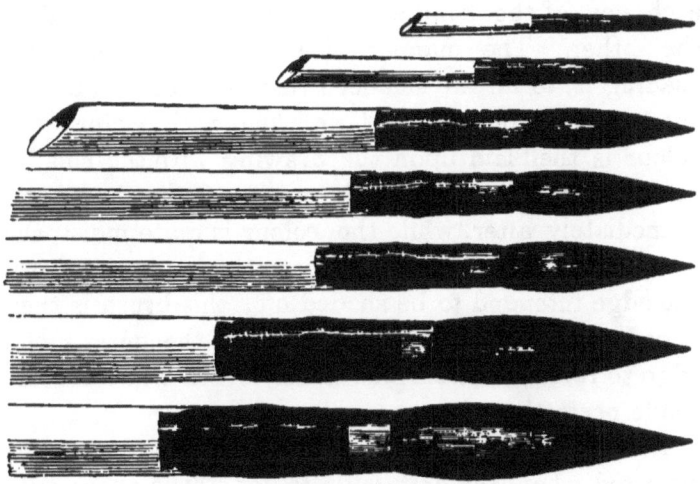

Sizes of Brushes.

The sizes of brushes are indicated by the quills in
which the hair is fixed: the above illustration gives
the complete series except the largest.   The names are
1 crow, 2 duck, 3 goose, 4 large goose, 5 small swan,

6 middle swan, 7 large swan.   The largest, the one not represented, is termed eagle.

In using the sable or other hair brush, it is always well to have as large a one as can safely be used, as the more colour a brush contains, the more equally will it deliver the required tint.   A large brush may and should have a very fine point.   For drawing in fine line ornaments, of course a very small brush is absolutely required.

Softener.

For shading, camel or sable hair brushes, called Softeners, are generally used: these have a brush at each end of the handle, one being much larger than the other.   The manner of using the softener for shading is, to fill the smaller brush with colour, and to thoroughly moisten the larger one with water; the colour is then laid upon the drawing with the smaller brush, to represent the dark portion of the shade, and immediately after, while the colour is quite moist, the brush that is moistened with water is drawn down the edge intended to be shaded off; this brush is then wiped upon a cloth and drawn down the outer moist edge to remove the surplus water, which will leave the shade perfectly soft.

If very dark shades are required, this has to be repeated when the first is quite dry.   The above only indicates the method; some practice will be required to shade nicely.   For rounded or hollow surfaces, particularly if for representing bright metal or for small work, shades may often be made very effective by having a few lines drawn over them with the drawing pen. In this manner also reflection on the dark edge drawn with the pen in Chinese white is often effective.

The front angles to the light of finished mechanical drawings, in elevation or perspective, are also much improved in appearance if the original fine ink line, on the light edge, is drawn over with a pen containing Chinese white. It takes off the hardness, and brings the angle forward.

To tint large surfaces, a large camel-hair brush is used, termed a Wash-brush. The manner of proceeding is, first, to tilt the drawing, if practicable, and commence by putting the colour on from the upper left-hand corner of the surface, taking short strokes the width of the brush along the top edge of the space to be coloured, immediately following with another line of similar strokes into the moist edge of the first line, and so on as far as required, removing the last surplus colour with a nearly dry brush. The theory of the above is, that you may perfectly unite wet colour to a moist edge, although you cannot to a dry edge without showing the juncture. For tinting surfaces, it is well always to mix more than sufficient colour at first. In sky effects two or three tints may be used; the better plan for applying these, if the drawing is otherwise finished, is to turn it upside-down, and to wash away from the work, changing tints as required. The last wet edge may then be brought to the margin of the paper to be cut off in removing the drawing from the board. For monochrome the sky should be done first; the lights may be left in the work sufficient to give effective roundness.

Skies and other wide surfaces may be perfectly tinted by wetting the whole surface of the paper first.

For blue and white skies. The blue will soften itself off into the wet surface sufficiently to give a cloud-like edge, and with a little practice will look very nice. The white may also be clouded with black or brown, while the paper is wet, in the same manner. The habit is soon attained by copying sky effect on this system.

In sketching landscapes and buildings I have found good perspective effect, and freedom from rawness in the picture may be obtained by tinting the whole surface before commencing the work. The pleasantest effect, and most true to nature, is attained by wetting the whole surface of the paper, and tinting it with the prismatic shades which fall naturally over sky and landscape. These follow in the following order : Blue in the zenith, shading off to yellow on the horizon, and advancing to red in the foreground, with such shifting of the quantities of the tints as we witness in changes of seasons and times of the day. For average effects I have found the following method best. Having wetted the whole sheet, draw a full brush of weak Indian yellow along the horizon, shade this off with a wet brush, or *badger*, if one is to hand, until the tint disappears at about one-fourth of the heights of the intended drawing from both the top and bottom edges. Now take a full brush, while the paper is still wet, of weak cobalt, and draw it two or three times along the top edge, and afterwards soften this off until it quite disappears near the horizon. Then take a full brush of weak scarlet lake, and soften this off in similar manner up to the horizon. The paper is now allowed to dry thoroughly, and when dry should present over the whole surface a prismatic shade. This may be done the day before it is intended to make the sketch or highly finished drawing. I have never found this plan of working appear wrong; the work on the tint appears generally more solid, and yet more aerial, than on white paper; and although the whole paper is tinted, any parts left untouched in the colouring appear clear white.

For landscapes and trees, which accompany architectural perspective, a very general defect, in the trees especially, is the excessive colour; the trees appearing often as green or brown masses. A very good way to avoid this is to sketch in the trees from nature, and to paint the colour exactly, or as nearly as the judgment may, as the colours appear to the eye at once, and not in any case tint over tint. For myself, I take the shades first; but this is indifferent. But the principle that I wish to observe particularly is, that of colouring no more than you see before you, and this with the colour that you see, to make sure of the natural breaks and lights through the foliage which add so great a charm to natural effects. When I make this observation, I am quite aware that the clear observation of actual colour is not universal with us, and some have to match up, as it were, until approximation is attained, or mix their colours by rule or instructions. But with many, I have no doubt, the observation of colour is more or less exact who do not make full use of it. If the habit of persevering with three pure colours only, say lake, Indian yellow, and indigo, for landscapes is persisted in, the mind, if the power of observing colours is only moderately developed, soon acquires the power of observing the quantity of the pure colours in every mixed tint of nature, and shortly the number of colours become rather an incumbrance than a help. This of course applies only to sketching from nature,—not to decorative or geometrical work or illuminations, where colours can scarcely be obtained of sufficiently pure tint to bear combination for brilliant work.

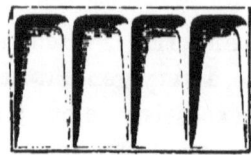

Slope Tile.

Water colours are ground by rubbing the colour with water upon some kind of palette; there are three kinds in general use by the profession. That termed a Slope, is a slab of porcelain divided into compartments, which slope down from the surface about half an inch; some of them slope less; but in the shallow ones, which are the cheapest, the colour dries very quickly.

Cabinet Nest.

A kind of palette, termed a Cabinet, which consists of a nest of six saucers, so constructed that the bottom of one forms the cover to another, may be recommended for general use, as by merely packing them together the colours may be kept perfectly moist during short intervals, as meal-times, etc.

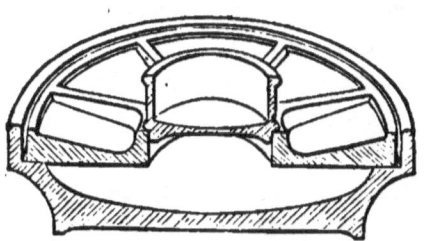

Wheel Slope and Basin, half-perspective section.

There is another kind of palette, of which the figure above is a half-perspective section; it consists of a circular slope with the centre space cut away; this rests over a kind of basin with a moveable cup standing up in the centre. This kind of palette is very convenient; the cup holds clean water, and the basin answers to wash the brushes in, either for changing the colour or before putting them away.

A new Ink Slab has recently been registered by Mr. Ackerman. In the centre of the saucer a small well is sunk, into which the ink naturally runs: a plug fits in the well which will expel the whole contents. The manner of using this palette is to pour in a little water, put the plug in the well to expel this, and then grind the ink as usual; when the plug is withdrawn the ink will run into the well, when it is out of use the plug is put in a hole over the well in the cover.

It is customary to use a water glass, which is a short kind of tumbler, with the two first-described kinds of palettes; with the last-mentioned it is unnecessary.

In conclusion, a few words may be said upon the very meritorious French chromo-lithographs, as examples of colouring.

In modern practice, the mere mechanical detail of geometrical drawing, which may be sufficient for prac-

tical purposes, is thought scarcely sufficient for the
perfect representation of objects that are required to
be in certain artistic proportions, however utilitarian
the purpose.   A steam engine, by the fitness of its parts
and excellence of workmanship, may be an elegant as
well as a useful machine ; in the same way, drawing,
by judicious but not heavy or elaborate colouring, will
appear much more pleasing, satisfactory, and comprehensible.   Therefore the art is very worthy of emula-
tion.   The French have long been considered better
colourists than the English, and perhaps there is
nothing would improve the young draughtsman equal
to copying a few of those beautifully coloured French
chromo-lithographs, which may be had in all the principal towns of this country, at very moderate prices.
For mechanical examples, certainly, nothing is equal to
them, and for architectural details they are very excellent.   If it is considered troublesome to copy them,
they may be kept and their colour effects applied with
advantage to produce excellent results.   They are
published in all styles of colouring—many of them too
heavy for English taste, but, generally speaking, so
varied, as to be by selection examples for all.

CHAPTER XXXV.

STENCIL PLATES, ETC.

STENCIL PLATES of late years have come into extensive use among draughtsmen : they effect an immense saving of time, and perform much of the most tedious and unremunerative labour.

The stencil plate is a thin sheet of copper or brass, perforated with letters or devices. By placing it upon a drawing, and brushing it over with ink or colour, the form of the perforations is delineated.

The perforations are made through the metal, either by engraving, by etching with acid, or, what is better, by both methods combined. If engraving only is employed, the force necessarily applied to the graver will sometimes stretch the plate unequally, whereas by etching alone, the edges of the perforations are left rough, and the corners imperfect; but if the line be lightly etched, and afterwards cleared with the graver, it may be rendered perfect without any risk of cockling the plate.

Copper is much better than brass for stencil plates; the metal being softer, it lies closer to the paper upon receiving the pressure of the stencilling brush. This close contact is a very important consideration, as it prevents the hairs of the brush from getting under the plate, and producing rough edges.

One of the most general purposes for which stencil plates are employed, is printing upon plans by means of alphabets and figures, which are made of various characters, the most used being the *block letter* shown last in the cut; this, having all the strokes of equal

thickness, is one of the most imperfect stencil letters, there being so many breaks which have to be left in the metal to give support to interior portions, as the centre part of O, B, D, etc.; thus to make block letters look sightly, it is necessary to fill up the breaks with the colour employed in stencilling. The letters which appear most perfect are shaded outline, old English, and ornamental. Although there are breaks in these, by the style or ornamentation they can scarcely be noticed.

The plain stencil alphabets will not be thought necessary to a draughtsman, if he be a good writer, as they will only save him a little time. A greater saving may be effected by the use of words which are constantly recurring, as Ground Plan, Front Elevation,

Section, etc.; or of interiors, as Drawing-room, Kitchen, etc. Of these a useful set may be purchased either for the architect, engineer, or surveyor, which will include all the words generally used.

For railways or public works, headings of plans may be cut especially, in suitable character and style; also words which are frequently repeated on any particular works, as the name and address of the architect or engineer, and any other constantly required words or sentences.

Besides letters and words, there are many devices by the use of which a superior effect may be produced, and much time saved; of these may be mentioned, north points, which are now very generally used; also plates for the representation of surface of country, as plantation, wood, marsh, etc., which, in drawing parish or estate plans, will save much labour; corners and borders for finished plans, and many other devices which will suggest themselves to the draughtsman.

Many persons fail in the successful use of stencil plates, either from the want of sufficient care or the knowledge of some particular requiring constant attention. The following observations may be useful. The brush requires to be squarely and equally cut, and to be kept moderately clean. If Indian ink is used, the largest surface of the cake should be taken to rub the moist brush upon, to get it equally diffused and softened with colour. A very cheap coarse kind of ink is generally sold with the stencil plates, which answers better than Indian ink, as it runs less upon the drawing and presents a larger surface to the brush.

After the plate has been in use some time, the fine lines and corners become clogged with ink; this may easily be removed by soaking the plate a short time in warm water, and afterwards lightly brushing it upon a flat surface until it is quite clean. It must be par-

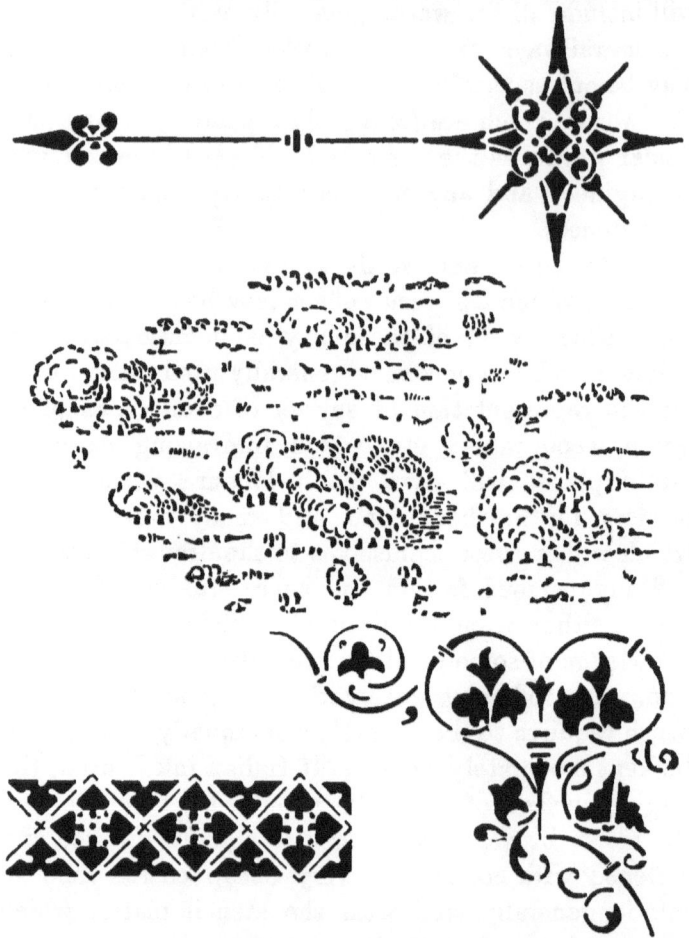

ticularly observed that a cloth should at no time be applied to the plate, either to clean or to wipe it, as this would be almost certain to catch in some of the perforations, and probably spoil the plate.

If the plate by improper use become cockled, it may be flattened, if laid upon a hard flat surface, by drawing a cylindrical piece of metal, as, for instance, the plain part of the stem of a poker, firmly across it several times on each side of the plate.

In using the stencil plate, it should be held firmly to the drawing by *one edge* only, in no instance allowing the fingers to cross to the opposite edge. The general method is to place the fingers of the left hand along the bottom edge. When the brush is diffused with ink, so that it is just moist, it should be lightly brushed upon a book-cover or pad, so as to free the points from any excess of colour. In applying the brush to the plate, it should be held quite upright, and moved, not too quickly, in small circles, using a constant, equal pressure, as light as appears necessary. The stencilling should be commenced at one end of the plate and proceeded with gradually to the other, moving onwards as the perforations appear filled with colour, being particularly careful not to shift the fingers placed upon the plate during the operation. If the plate is very long the fingers may be shifted after each word, if the plate be held down during the time firmly by the other hand. Should there not be quite sufficient ink in the brush to complete the device, the plate may be breathed upon, which will moisten the ink attached to it. If, after the plate is removed, the device appear light in parts, the plate may be replaced and the defects remedied, if very great care be taken to observe that the previous stencilling perfectly covers the perforations.

In stencilling words or numbers with the separate letters of the alphabet, a line should be drawn where the bottoms of the letters are intended to come. The separate letters should be taken as required and placed upon the line, so that the line just appears in the perforations. That the letters should be upright, it is best that the next letter on the slip used should also allow the line to appear in it in like manner. The required distance of the letters apart must be judged of by the eye, a pencil mark being made after each letter is completed to appear in the perforation on the near side of the next letter to be stencilled.

With care, a stencil plate will last in constant use for many years; without care, it is practically spoilt by taking the first impression.

INDEX.

PRICE LIST

OF THE

INSTRUMENTS DESCRIBED,

MADE BY THE AUTHOR,

And Sold at 4 & 5, Great Turnstile, Holborn, London, W.C.

FIRST QUALITY.*

Page				£	s.	d.
8	Drawing Pens, fine steel . . . . . each			0	2	6
9	,,	block or detail . . . . ,,		0	3	6
9	,,	lithographic . . . . . ,,		0	4	0
10	,,	lifting-nib, electrum . . . ,,		0	5	0
10	,,	Stanley's improved, electrum . . ,,		0	7	0
10	,,	ditto, nib not to lift, steel . . ,,		0	3	0
11	,,	solid ,, . ,,		0	6	0
11	,,	Stanley's curve ,, . . ,,		0	4	6
13	,,	lithographic crowquill, per ¼ doz., 1s., per doz.		0	2	0
14	,,	plain bordering ,, . . each		0	4	6
14	,,	Stanley's border or colour, elec. or steel ,,		0	6	6
15	,,	road or double, steel . . . ,,		0	10	6
15	,,	section ,, . . . ,,		0	6	0
16	Road pencil ,, . . . ,,			0	10	6
17	Improved wheel pen, with set of four wheels, electrum ,,			0	8	6
19	Tracer, steel or agate . . . . . . ,,			0	3	0
19	Pricker, plain, with reserve, electrum . . . ,,			0	2	6
20	,,	patent ,, . . . ,,		0	3	6
21	Dividers, plain sector ,, . . . pair			0	5	6
22	,,	steel joint ,, . . . ,,		0	3	6
25	,,	hair ,, . . . ,,		0	8	0
26	,,	Stanley's improved ,, . . . ,,		0	9	0
26	,,	sheath, with scales ,, . . . ,,		0	15	0
27	,,	pillar and Napier ,, . . . ,,		0	15	0

* 2nd and 3rd qualities of many of the articles at much lower prices.

Engine-dividing to Scientific Instruments, both straight-line and circular, on Metal, Glass, Wood, etc.   x

CASES OF MATHEMATICAL DRAWING INSTRUMENTS.

£35.— Thirteen-inch solid oak case, with mediæval bindings, trays, drawer, etc., fitted up in the best manner, containing electrum instruments of first quality and finish. Portable beam compasses, 6-inch compasses, with two lengthening bars; 4½-inch compasses,—these are fitted with patented needle points, exchangeable for spring points, and with ink and pencil points. Ink and pencil bows, patent points, and spring bows with needle points. Triangular compass, with a sliding point to produce ovals in pencil; 9-inch proportional compasses with adjustment; 5-inch plain and 4-inch hair dividers; dotting pen, road pen, four drawing pens, crow-quill pens; road pencil, pricker, tracer, brick gauges, horn centres, pins, etc.; 6-inch circular protractor, with vernier and arm; 12-inch solid rolling parallel rule; seven 12-inch ivory scales and offsets, set squares, lettering squares, curves, etc. The drawer contains colours, ink, sable brushes, palette, etc. This is a most elegant and useful case, very suitable for presentation.

£24.—Thirteen-inch electrum-bound walnut-wood case, lined with best silk velvet, with two trays, Hobbs' lock, drawer, etc., containing the following extra finished instruments:—6-inch compasses with patent *A* and *B* points, reversible ink and pencil points, and two lengthening bars; patent pointed beam compasses, with ink and pencil points and adjustment; 9-inch engine-divided proportional compasses; triangular compasses; 4½-inch compasses, with patent point, ink and pencil points; 5-inch sector divider; 4-inch hair divider; ink and pencil bows, with patent points; set of three spring bows, to hold needles; improved dotting pen, with set of wheels; four drawing pens, road pen, pricker, tracer, knife key and other keys; 6-inch circular protractor, six 12-inch ivory scales and offsets; 12-inch solid electrum rolling parallel; angles and curves. The drawer contains ten cakes of colours, Indian ink, sable brushes, and palette.

£13 10s.—Thirteen-inch walnut-wood case, Hobbs' lock, lined with silk velvet, containing the following electrum instruments of the highest finish:—6-inch compasses, with patented *B* points, lengthening bar, ink and pencil points; improved hair divider; ink and pencil bows with patented *B* points; three best spring bows, plain points; beam compass heads with adjustment, ink and pencil points; proportional compasses, engine-divided; one improved drawing pen and two fine steel; pricker, tracer, dotting pen, horn centres, drawing pins, knife key; six 12-inch ivory scales (either architects' or chain); 12-inch vulcanite rolling parallel rule; angles, pear-wood curves, and horn protractor.

£9.—Square walnut-wood case, electrum bound, lined with silk velvet, tumbler lock, etc., containing the following extra finished electrum instruments :—6-inch compasses, with patented B points to hold needles, ink, and pencil points, and lengthening bar ; improved hair divider ; bows, patent points ; set of three spring bows, points to hold needles ; engine-divided proportional compasses, two drawing pens, pricker and knife key ; either a set of three architects' or engineers' ivory scales, or sector, protractor, and rolling parallel.

£5.—Morocco leather case, containing electrum 4½-inch double-jointed compasses with points to hold needles, with lengthening bar, ink and pencil points, pair of double-jointed ink and pencil bows to hold needles, set of three spring bows, hair divider, two pens, pricker, knife key, and ivory protractor.

£3 5s.—Rosewood case lined with velvet, containing the following electrum instruments :—6-inch sector-joint compasses, with ink and pencil points, and lengthening bar; ink and pencil bows ; hair divider; two drawing pens; pricker ; knife key; set of three ivory scales.

£2 2s.—Mahogany case, lock, etc., containing brass sector-joint compasses, with ink and pencil points, and lengthening bar ; ink and pencil bows, sector divider, drawing pen and pricker, ivory protractor scale, parallel, and box sector.

£1 5s.—Mahogany case, lock and key, shifting tray, containing 6-inch brass double steel-joint compasses, ink and pencil points ; divider, ink and pencil bows, drawing pen, set of three boxwood scales.

One hundred varieties of cases in stock.   Cases fitted to own selection on the shortest notice.

A complete Mathematical Catalogue, containing prices of Theodolites, Levels, Mining Dials, Sextants, Prismatic Compasses, Optical Squares, Levelling Staves, Chains, Tapes, Rods, Rules, etc., will be sent free by post on application,   Address, to Mathematical Department—

W. F. STANLEY,

4 & 5, GREAT TURNSTILE, HOLBORN, LONDON, W.C.

An Optical Catalogue, containing prices of Microscopes, Telescopes, Binocular Marine and Opera Glasses, Magnifying Glasses, Spectacles ; Barometers, Thermometers, Hydrometers, etc., etc., will be sent free by post. Also a Catalogue of Photographical Instruments and Chemicals. Address to Optical Department—

W. F. STANLEY, 13, Railway Approach, London Bridge, S.E.

www.ingramcontent.com/pod-product-compliance
Lightning Source LLC
Chambersburg PA
CBHW060534030726
47498CB00004B/1193